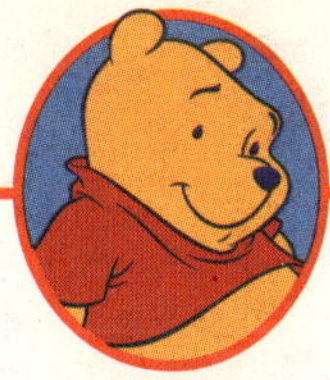

Colors
Table of Contents

McGraw-Hill
Children's Publishing
A Division of The ***McGraw-Hill*** *Companies*

Send all inquiries to:
McGraw-Hill Children's Publishing
8787 Orion Place
Columbus, OH 43240-4027

ISBN 1-56189-217-3

1 2 3 4 5 6 7 8 9 10 CJK 07 06 05 04 03 02

Name: ____________________

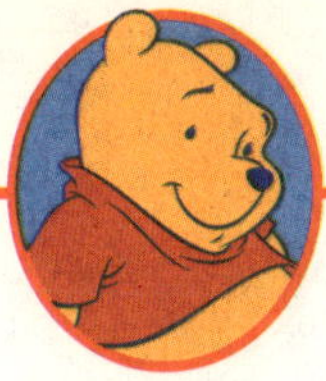

Red

Color the ladybug **red**.

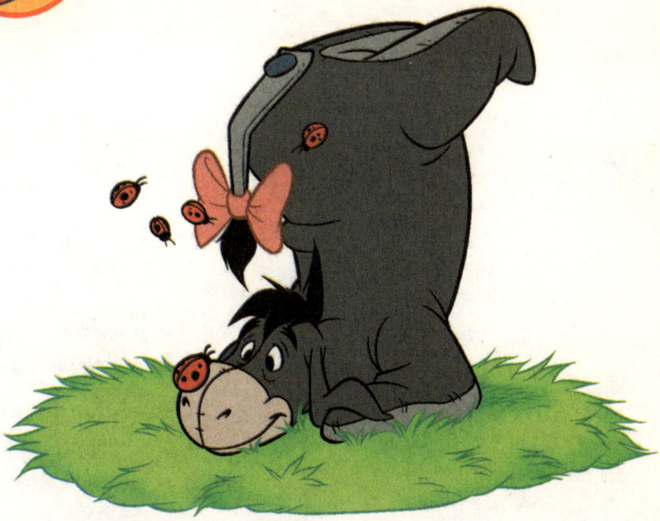

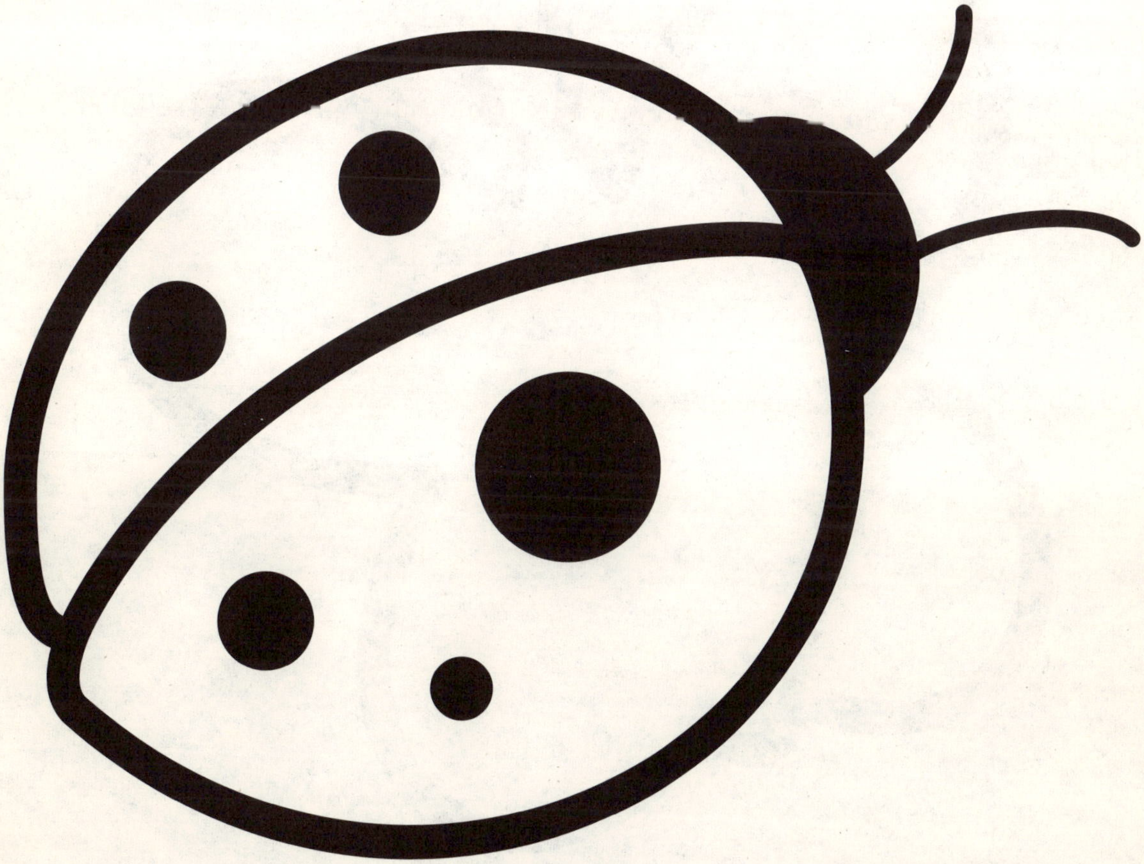

Name: ______________________

Blue

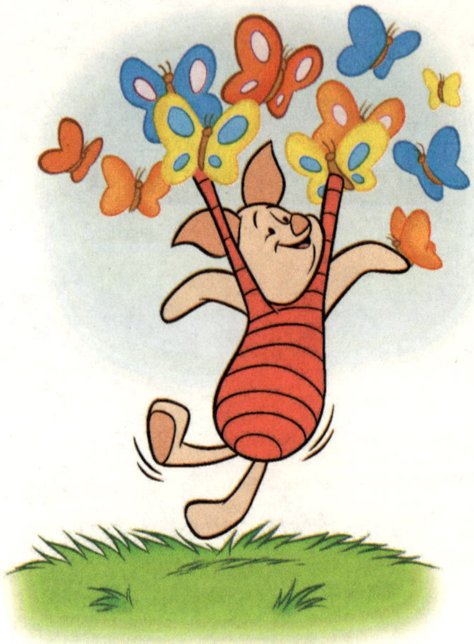

Color the butterfly **blue**.

Name: ____________________

Green

Color the leaves **green**.

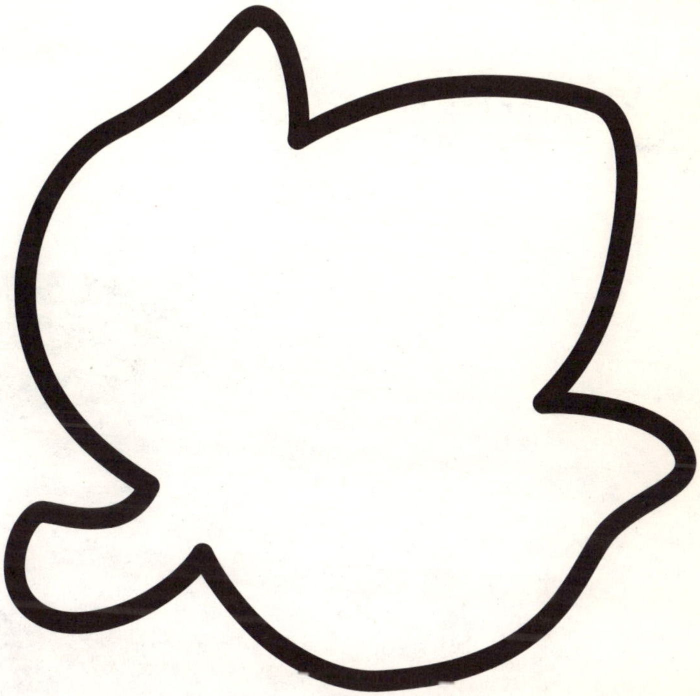

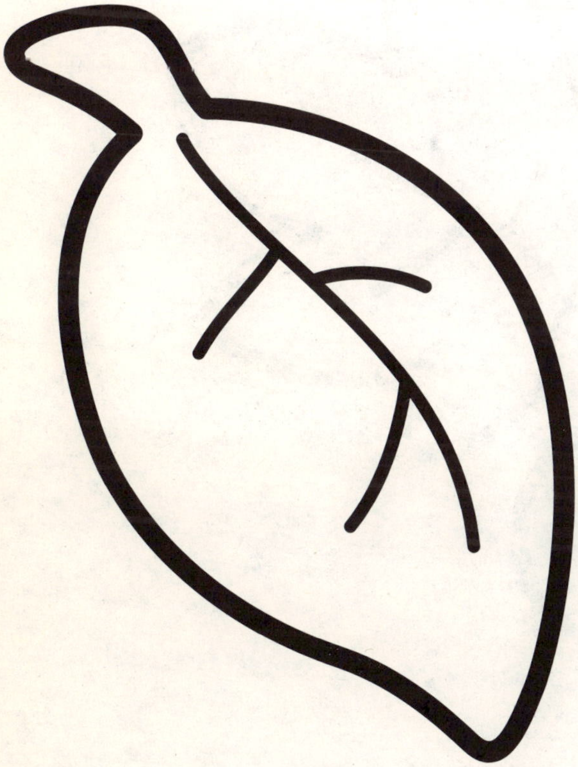

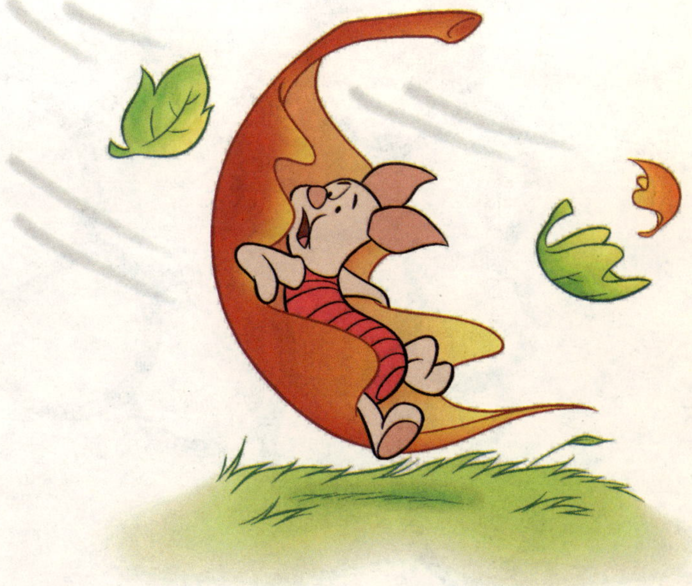

Name: ____________________

Purple

Color the grapes **purple**.

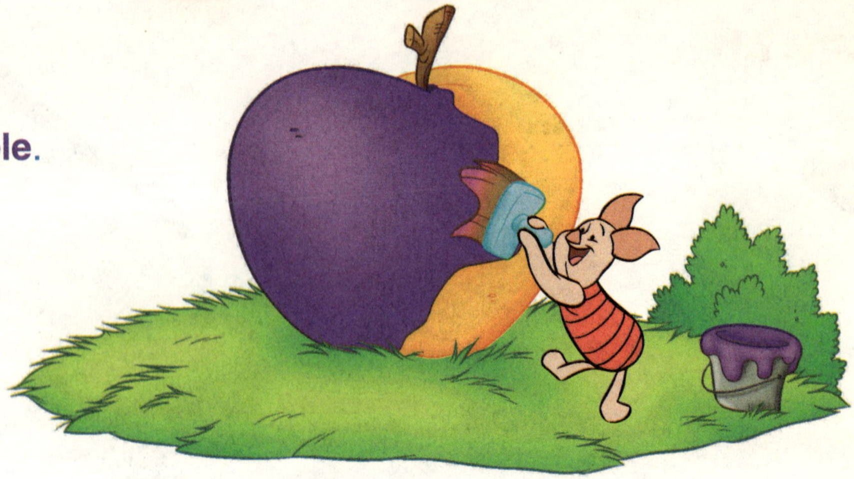

Name: ____________________

Yellow

Color the bees yellow.

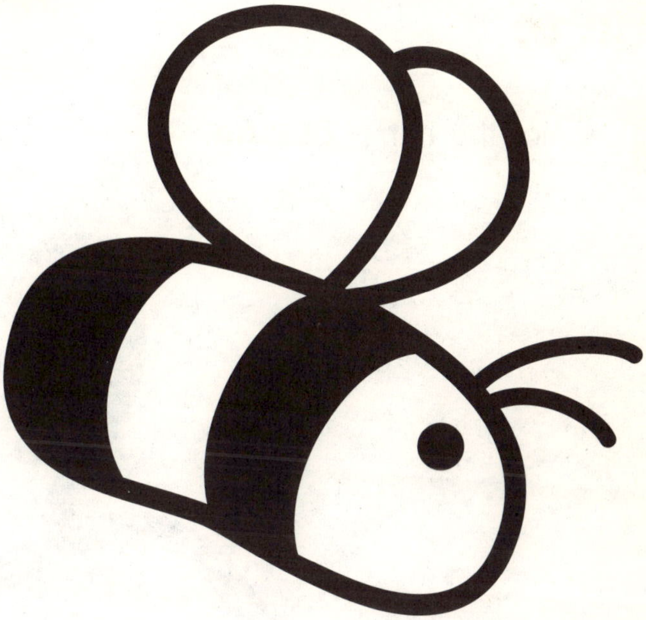

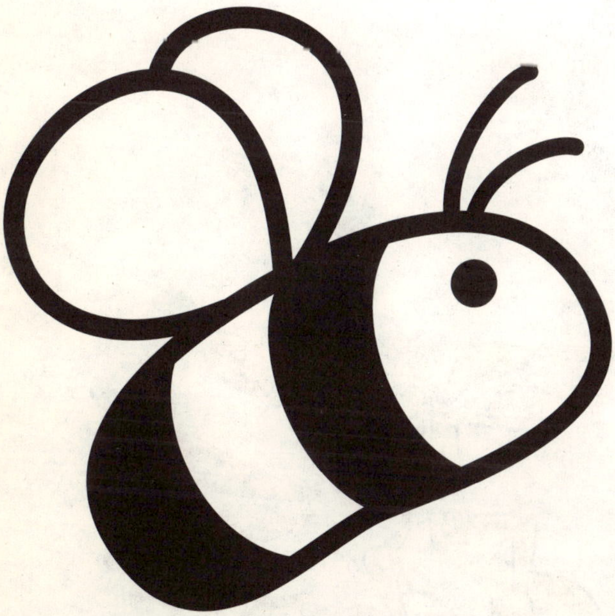

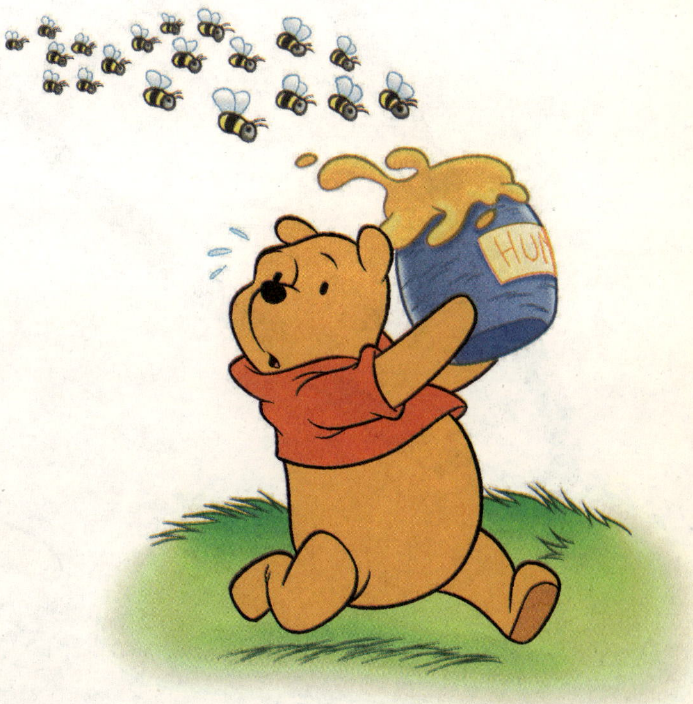

Name: ____________________

Orange

Color the carrot **orange**.

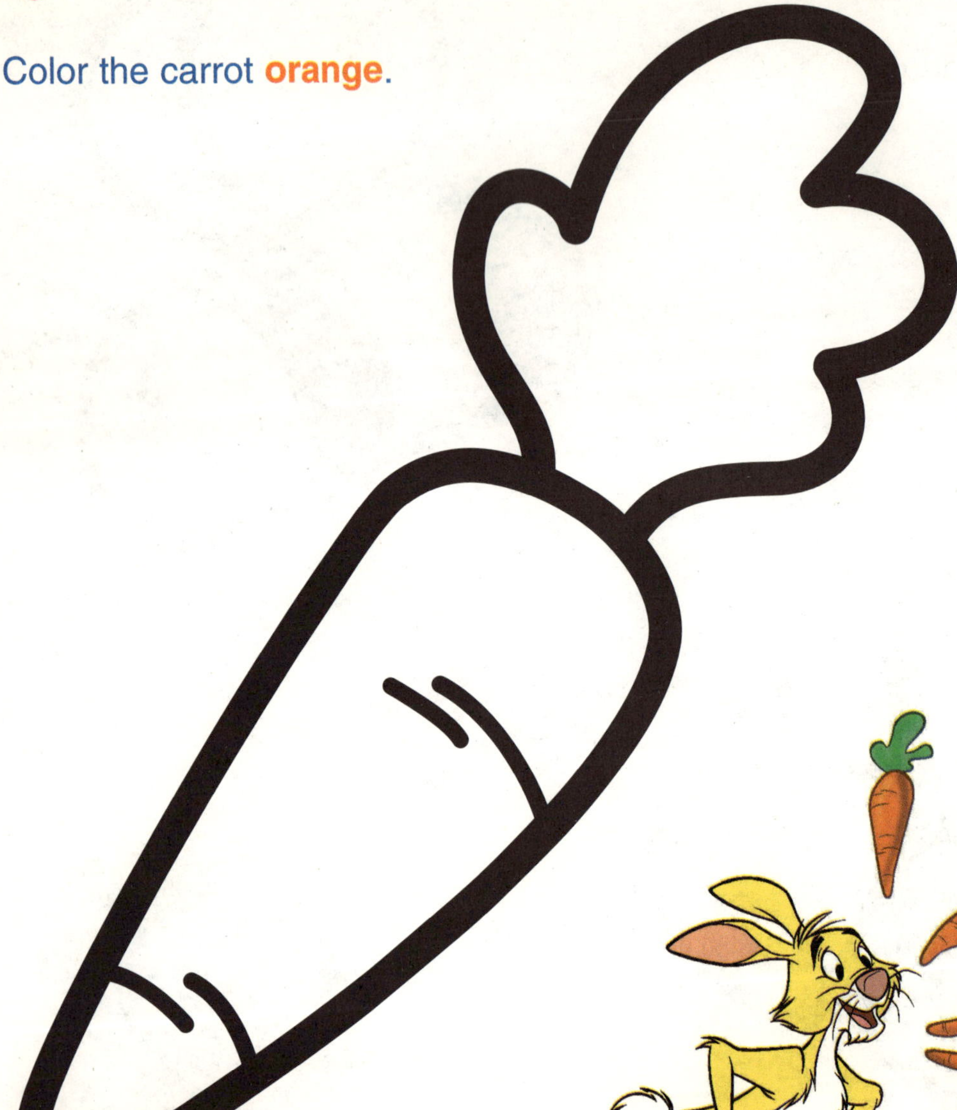

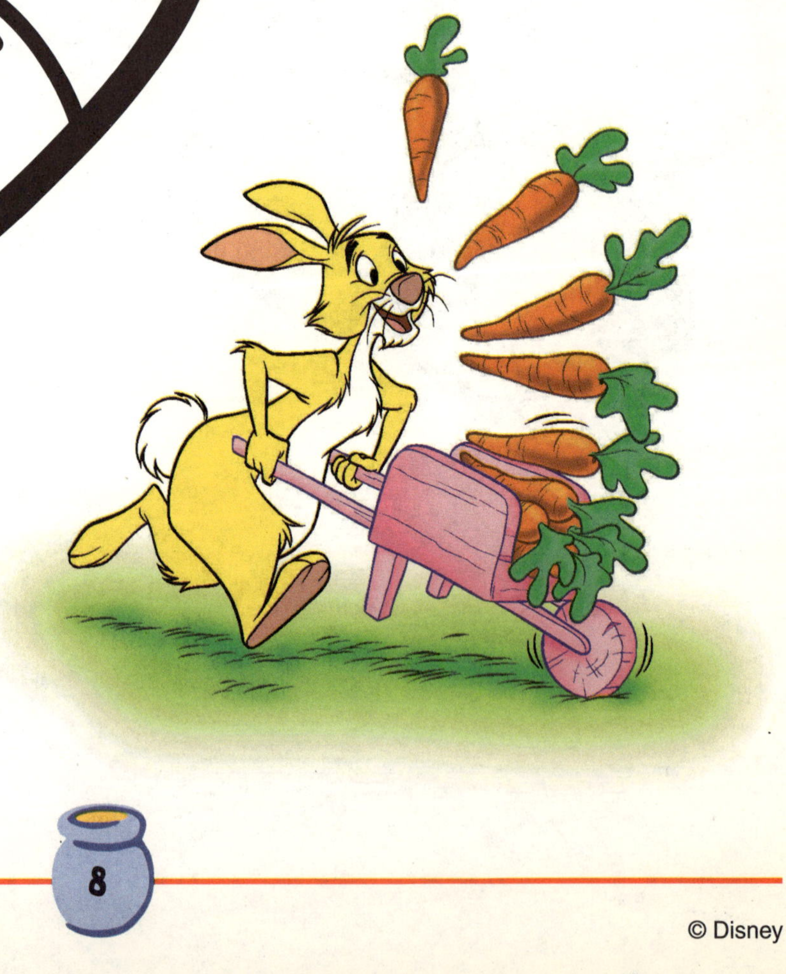

Name: ______________________

Brown

Color the acorn and paint bucket **brown**.

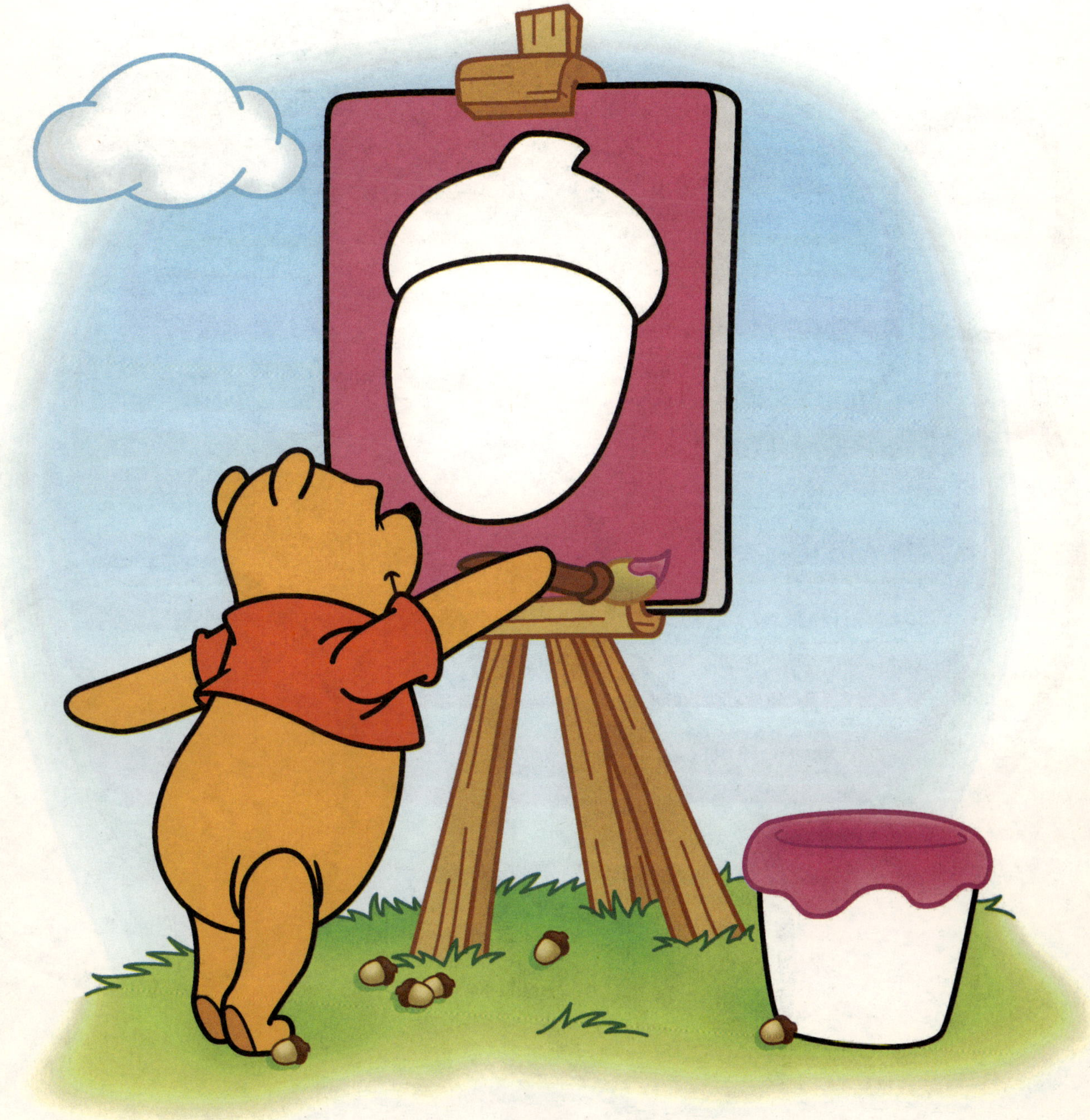

Name: ____________________

Black and White

Color the pins white.
Color the ball **black**.

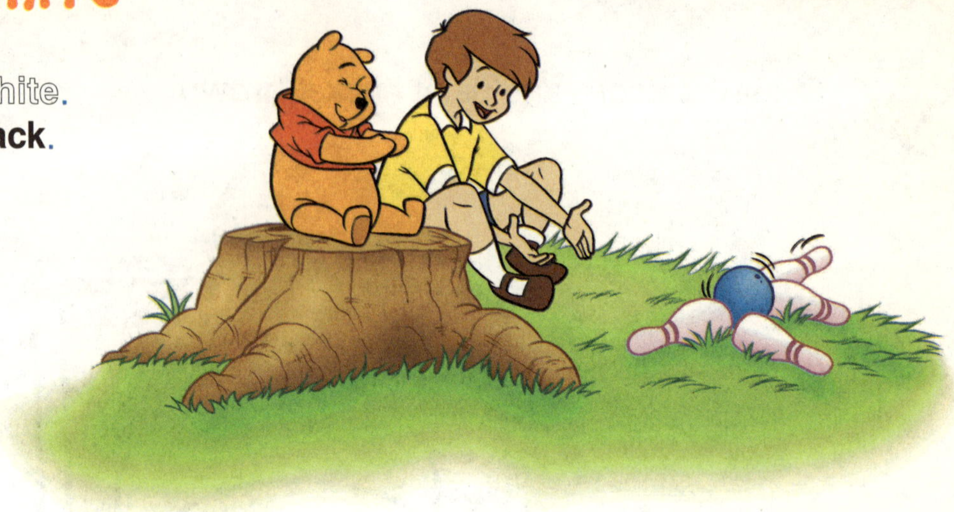

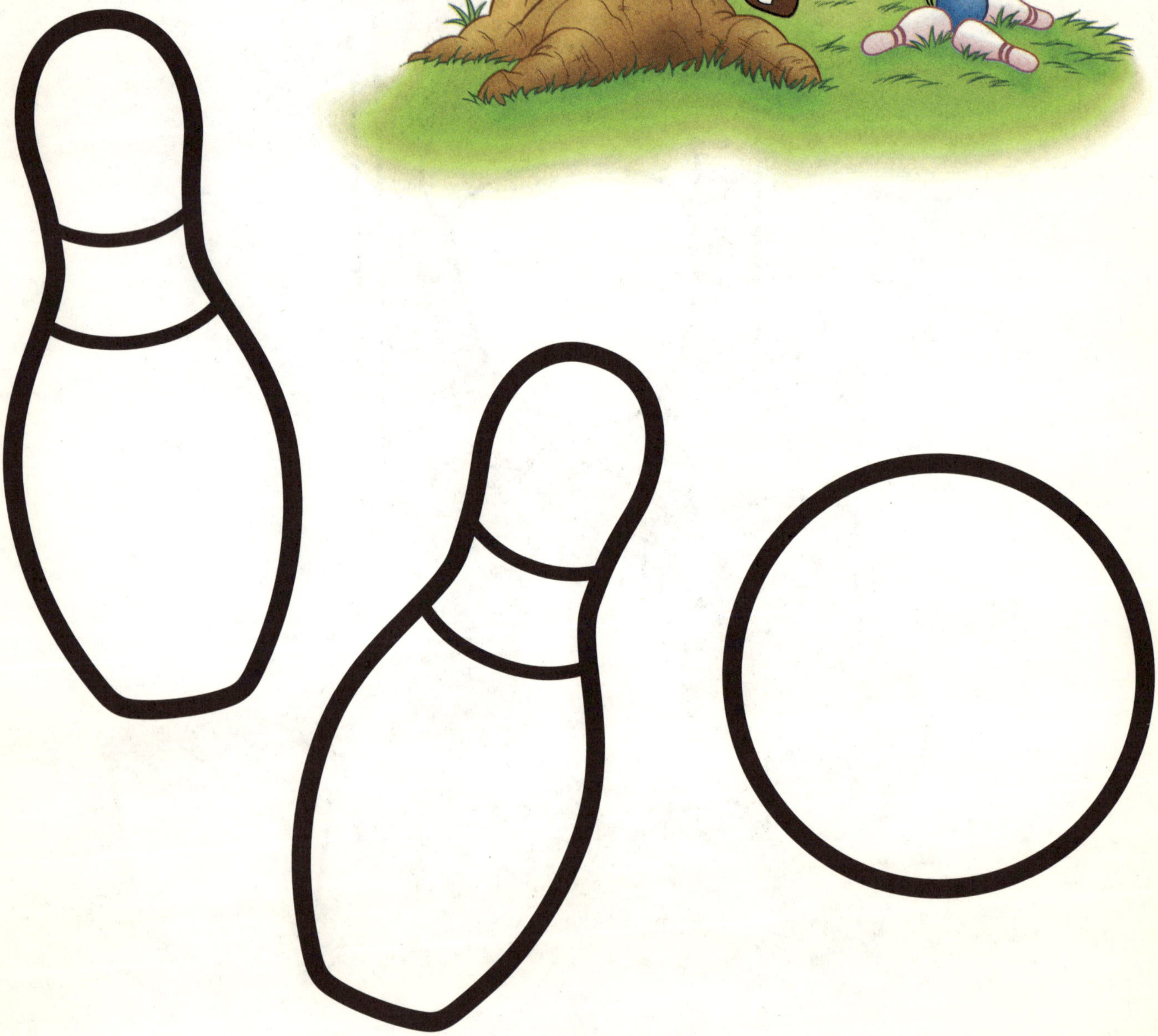

Name: ____________________

Colors Review

Help Piglet. Trace the balloons with the correct colors. Then, color them in.

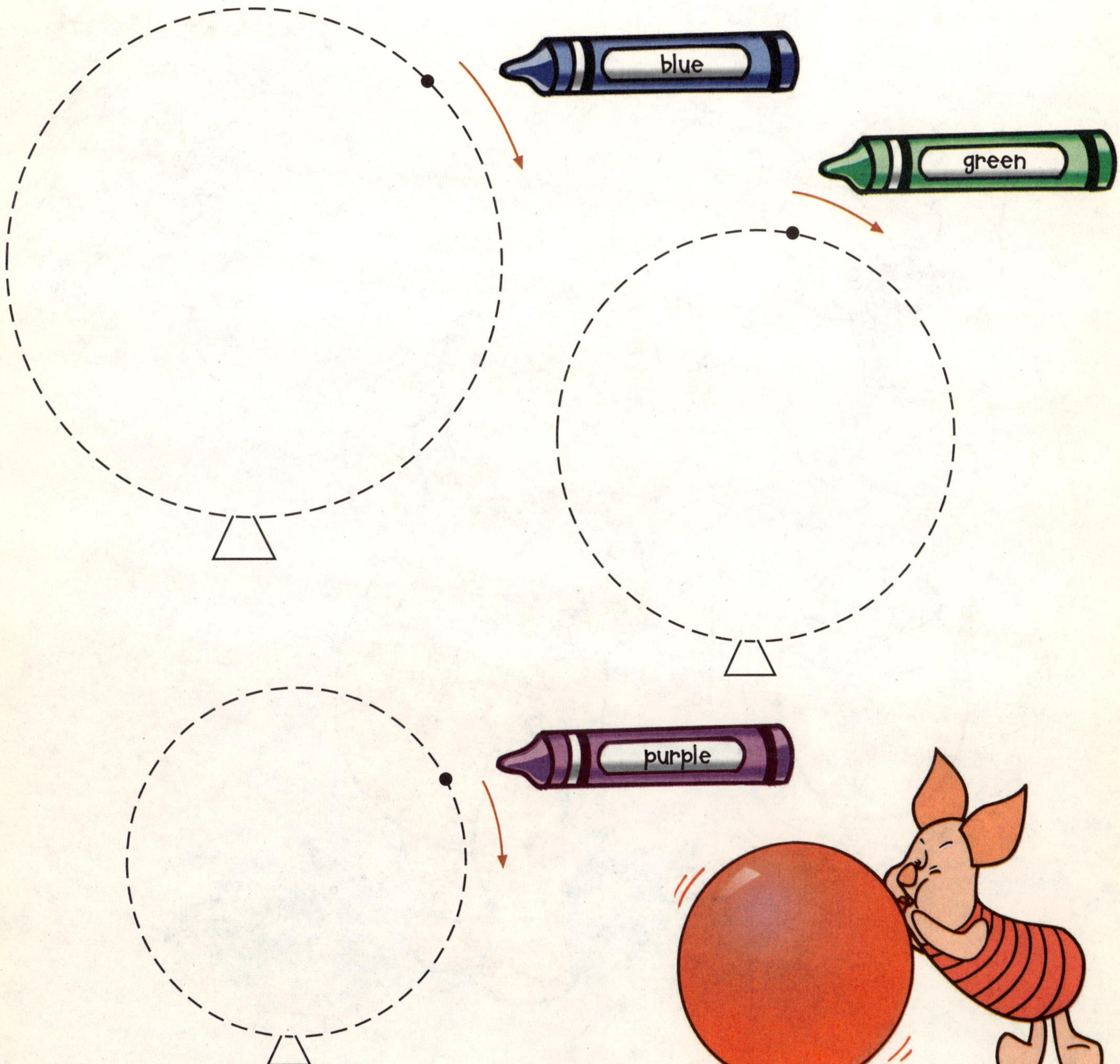

Name: ____________________

Colors Review

Read and color Pooh's thoughts the correct colors.

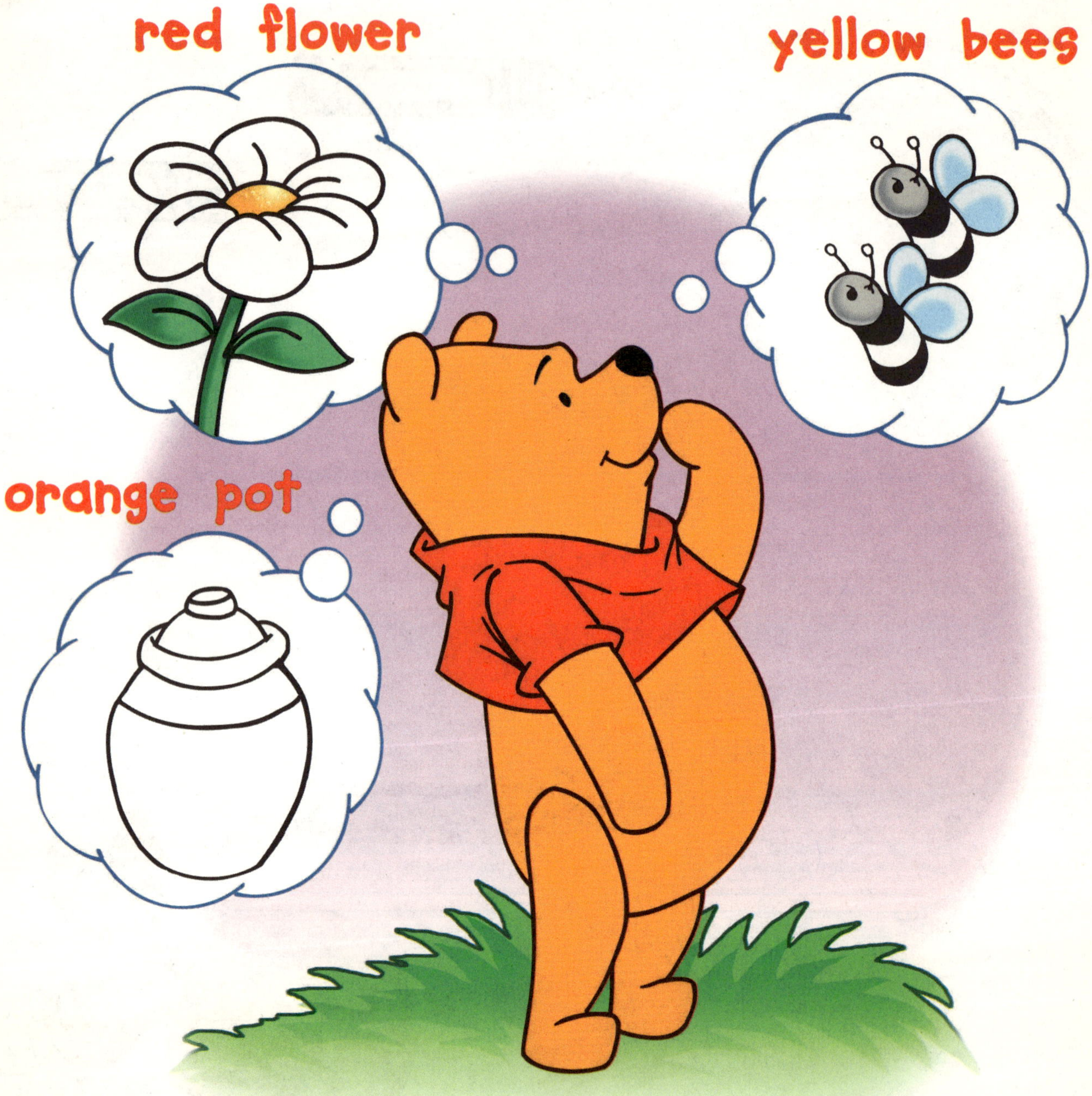

Name: ______________________

Colors Review

Read and color Eeyore's thoughts the correct colors.

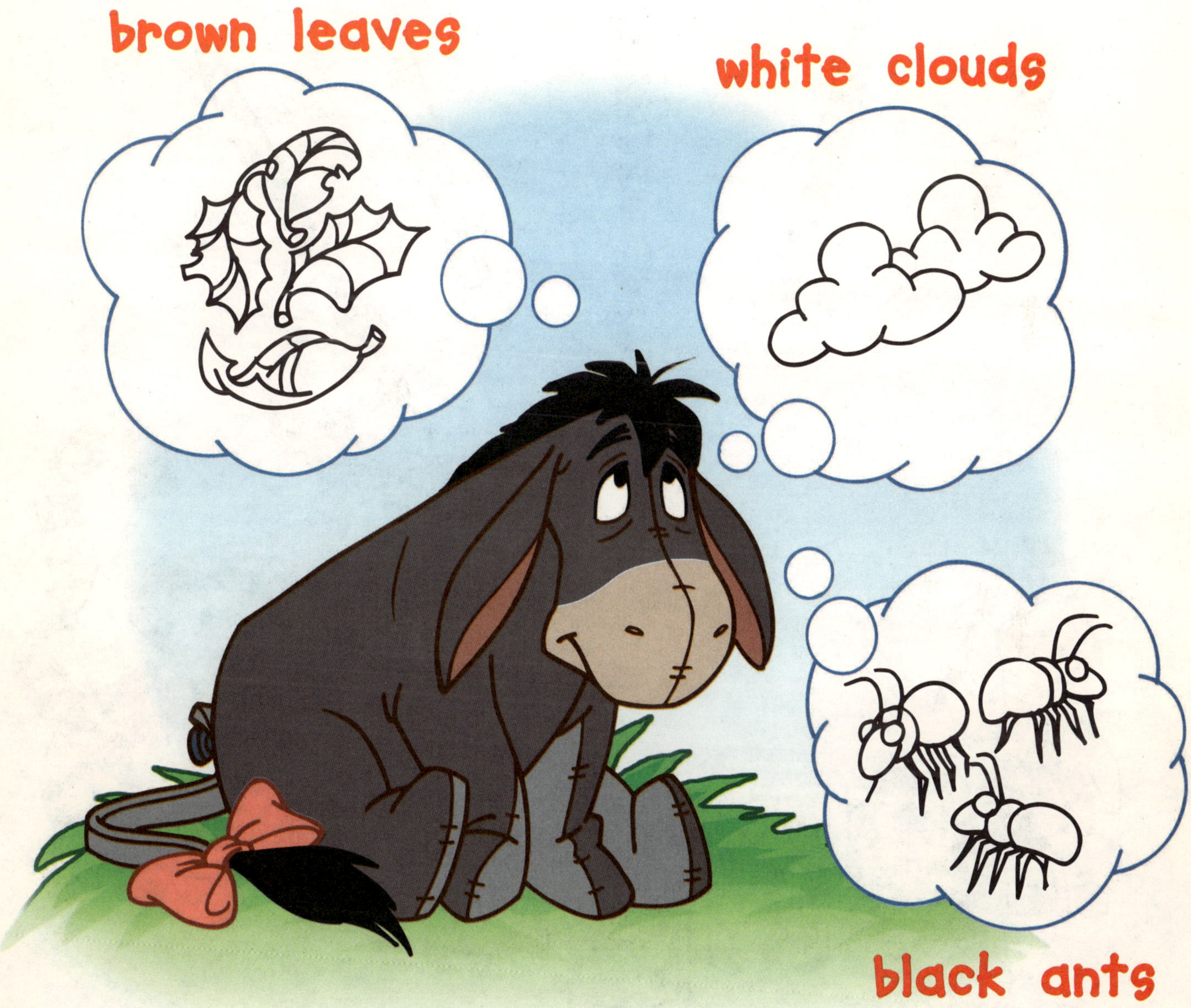

Name: ______________________

Matching

Help Tigger and Piglet. Draw a line to match each paintbrush with the same color paint can.

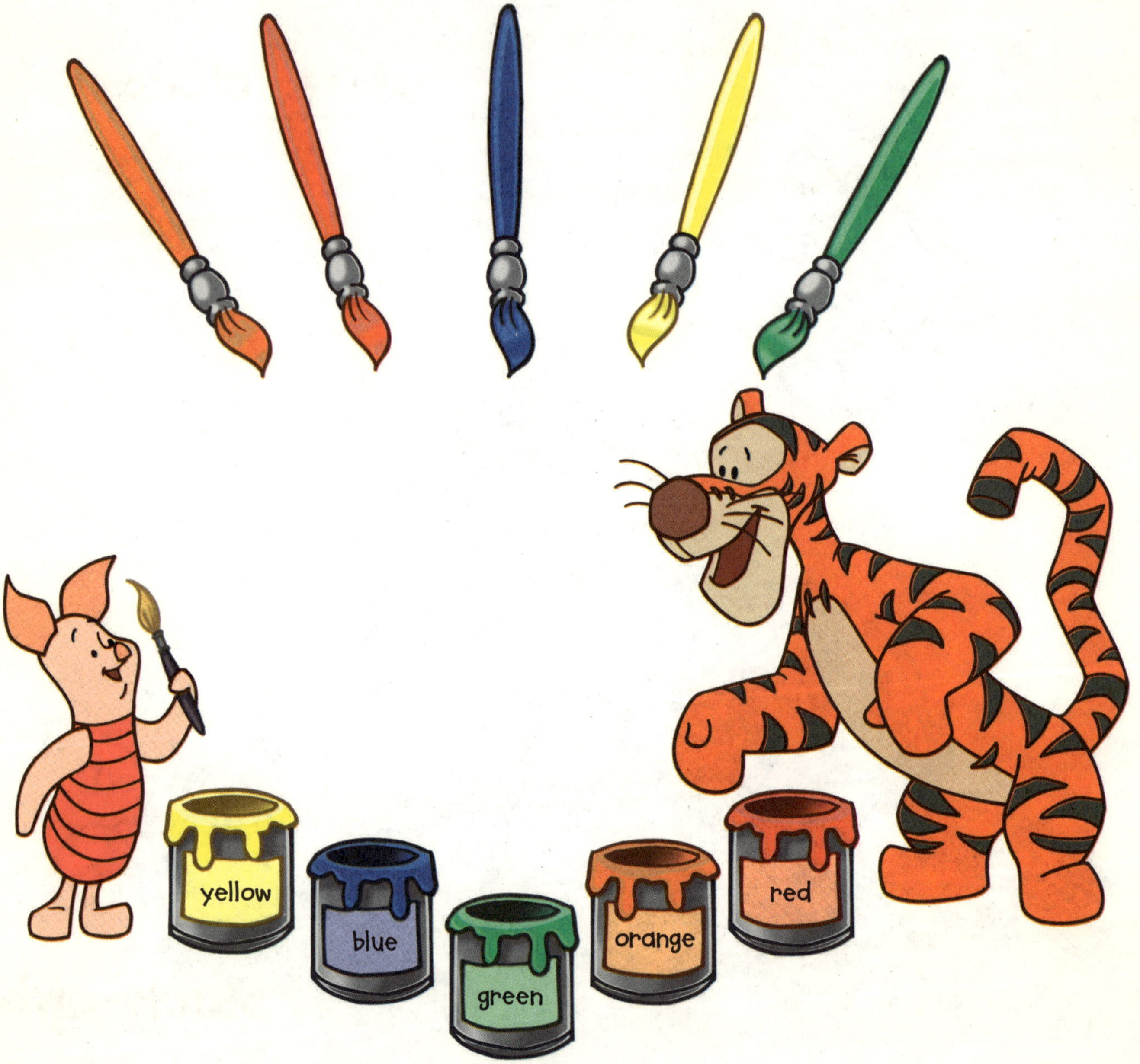

Name: ______________________

Reasoning

Do what Roo does. Color each drawing with the correct color.

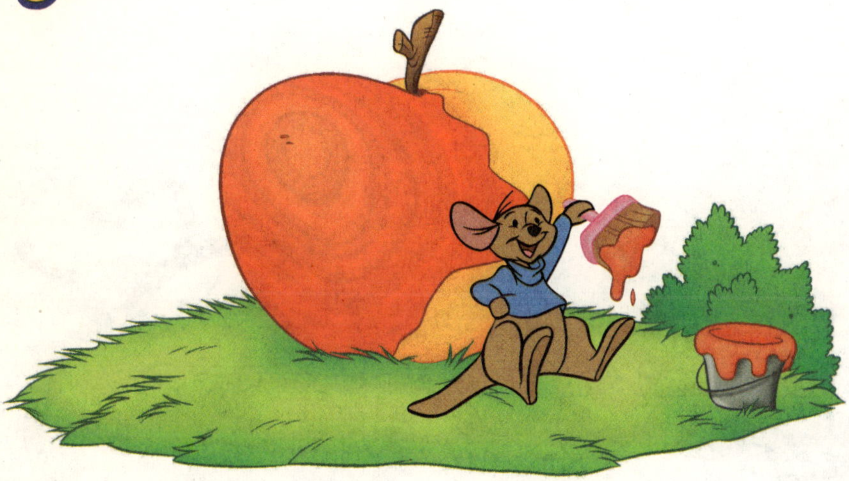

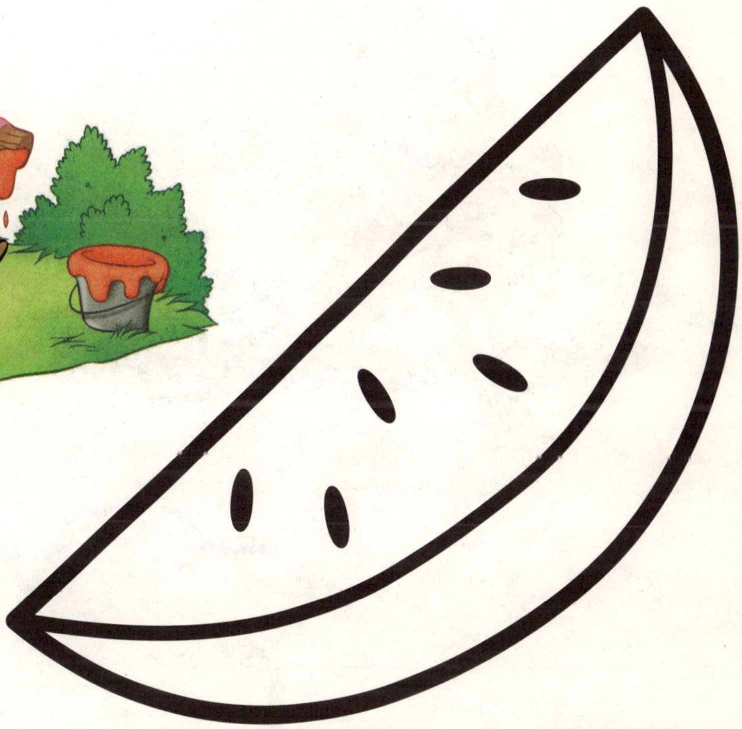

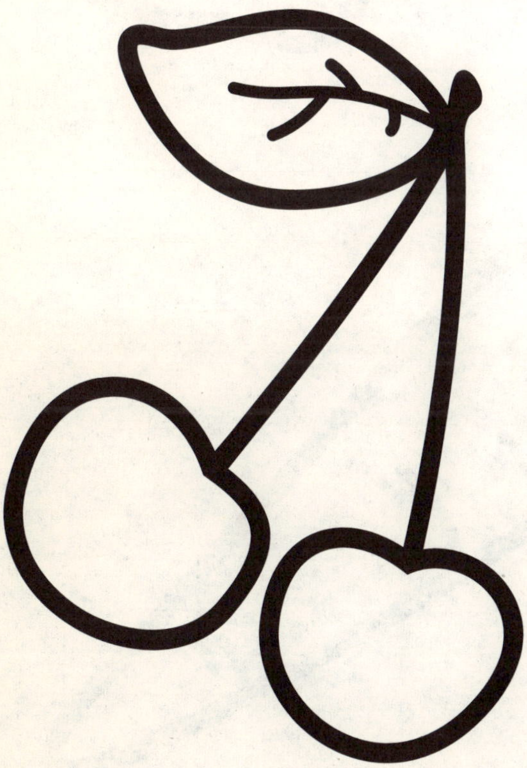

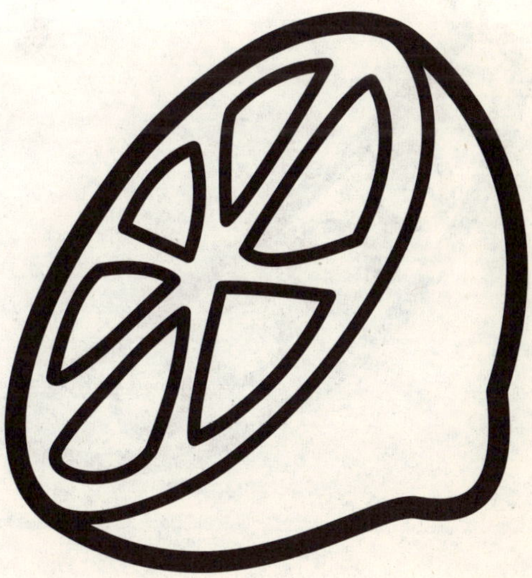

Name: ____________________

Reasoning

Do what Pooh does. Color each drawing with the correct color.

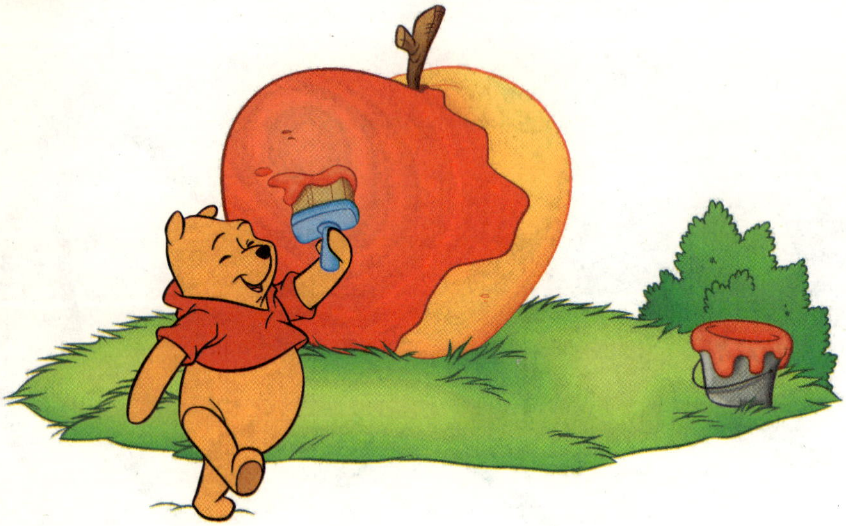

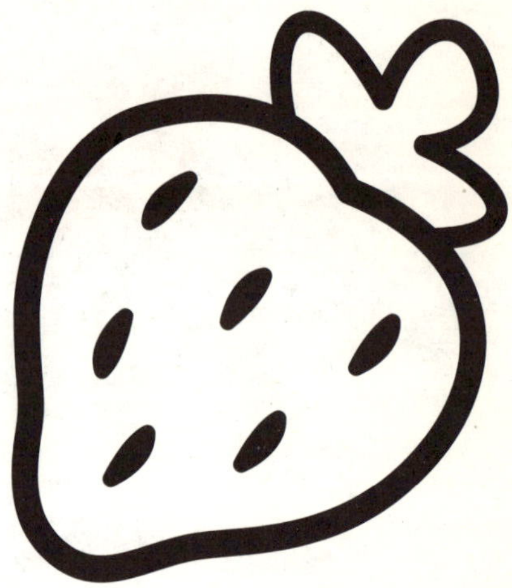

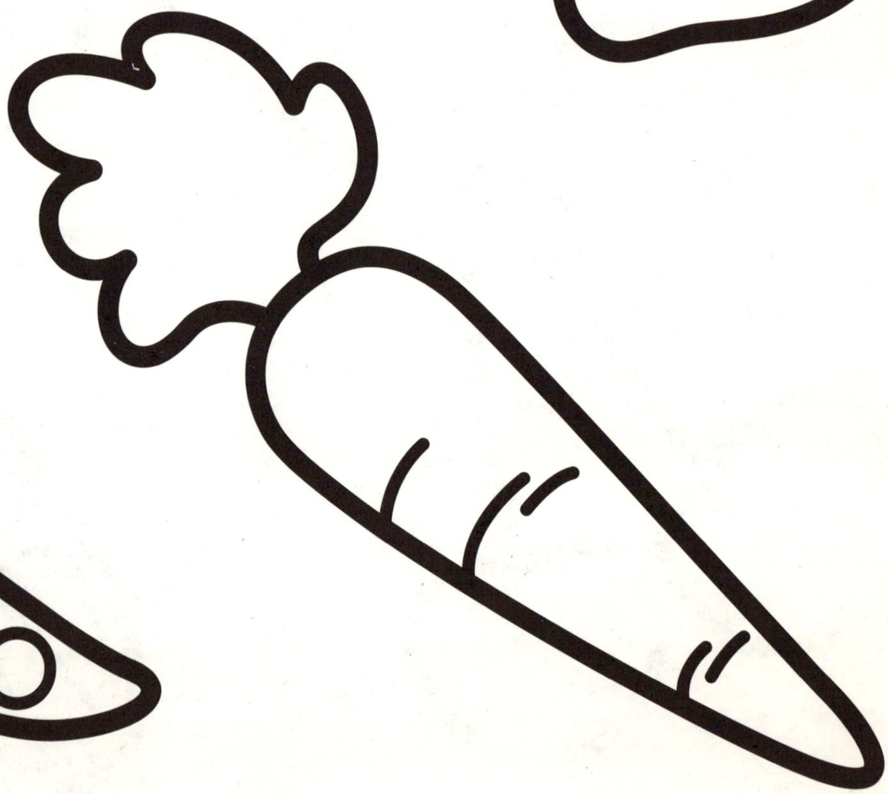

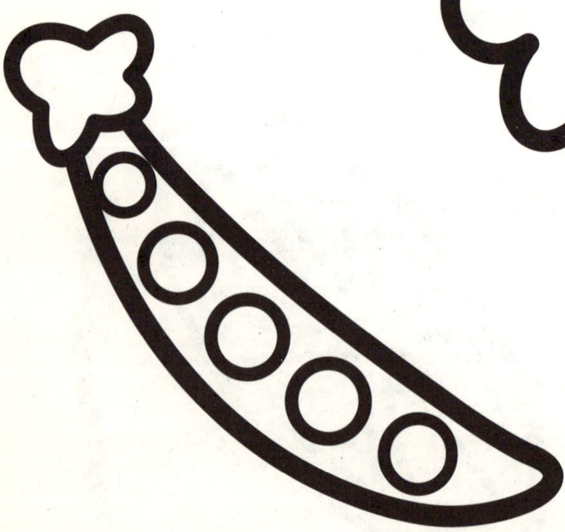

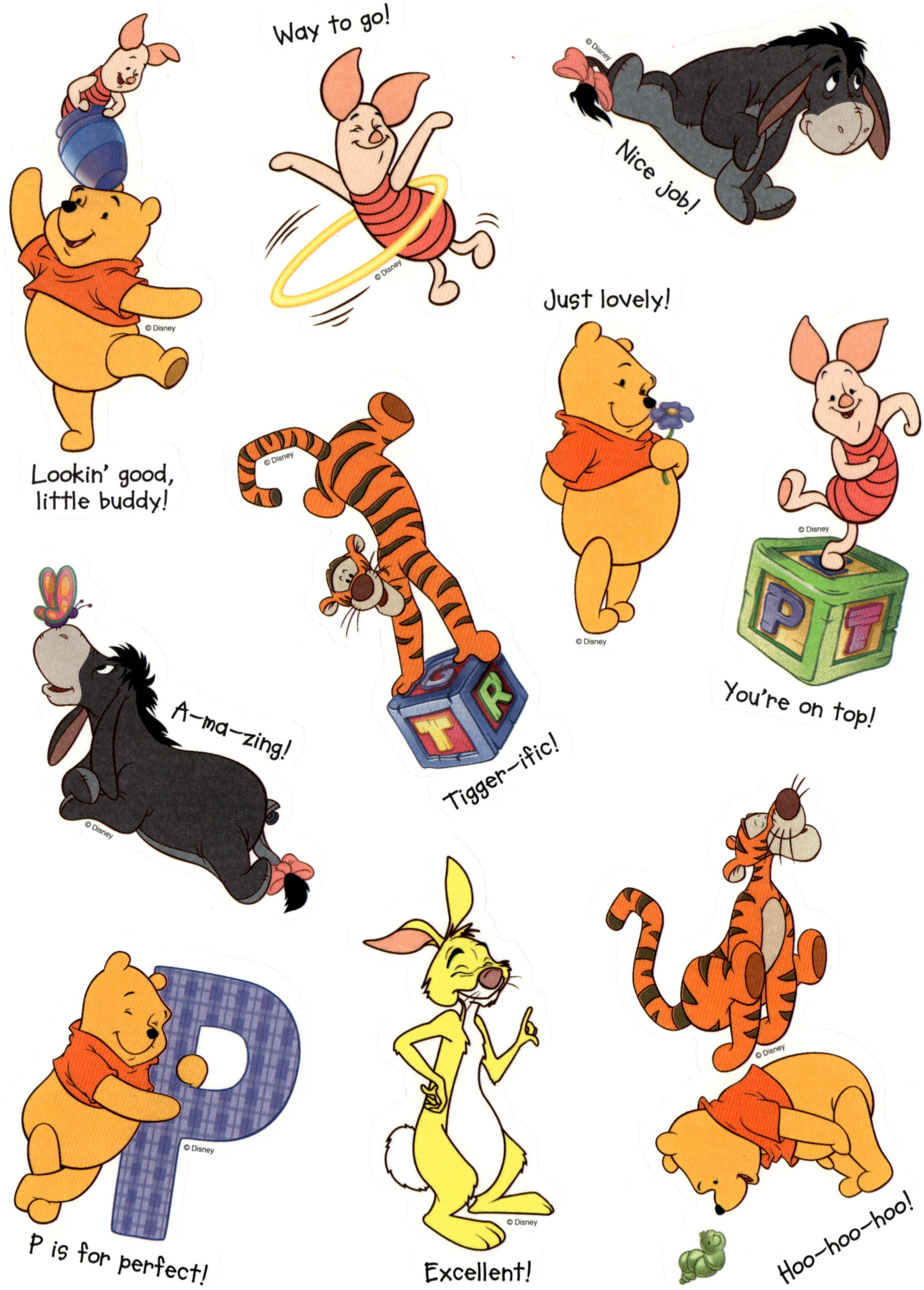
Way to go!
Nice job!
Lookin' good, little buddy!
Just lovely!
Tigger-ific!
You're on top!
A-ma-zing!
P is for perfect!
Excellent!
Hoo-hoo-hoo!
© Disney

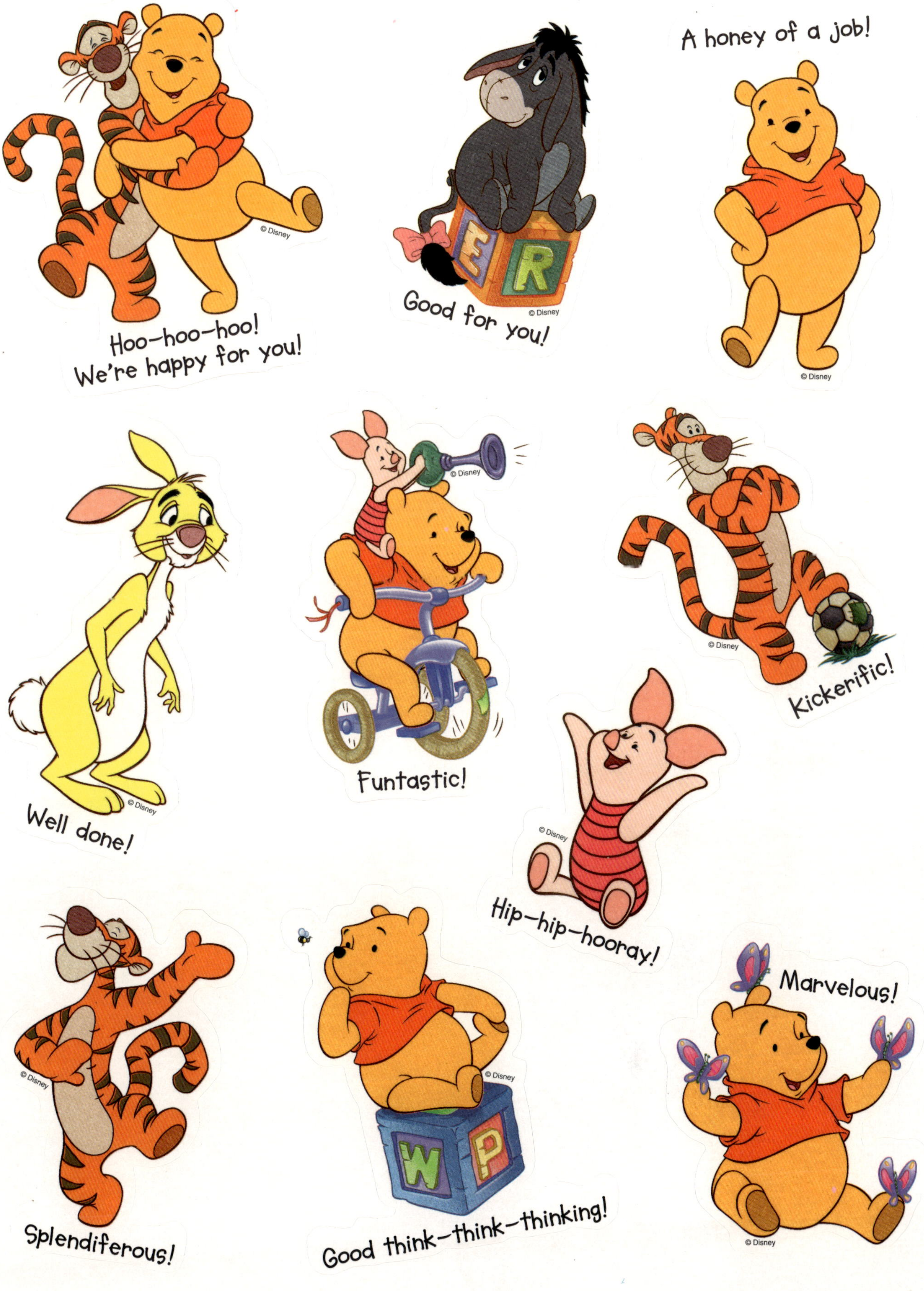
Hoo-hoo-hoo!
We're happy for you!
© Disney
Good for you!
© Disney
A honey of a job!
© Disney
Well done!
© Disney
Funtastic!
© Disney
Kickerific!
© Disney
Hip-hip-hooray!
© Disney
Splendiferous!
© Disney
Good think-think-thinking!
© Disney
Marvelous!
© Disney

Name: ______________________

Color Cues

Color the drawing with the same colors as the dots.

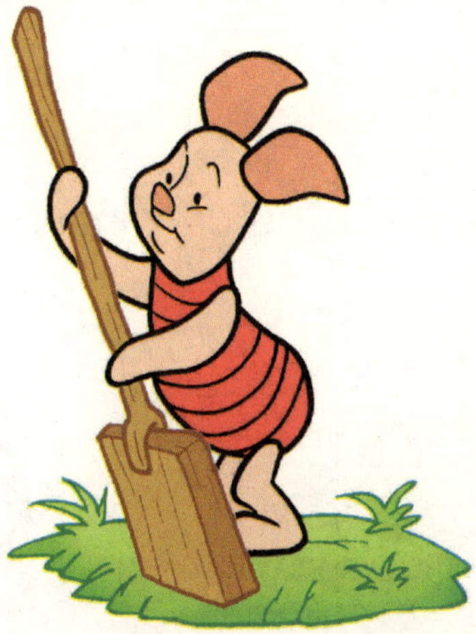

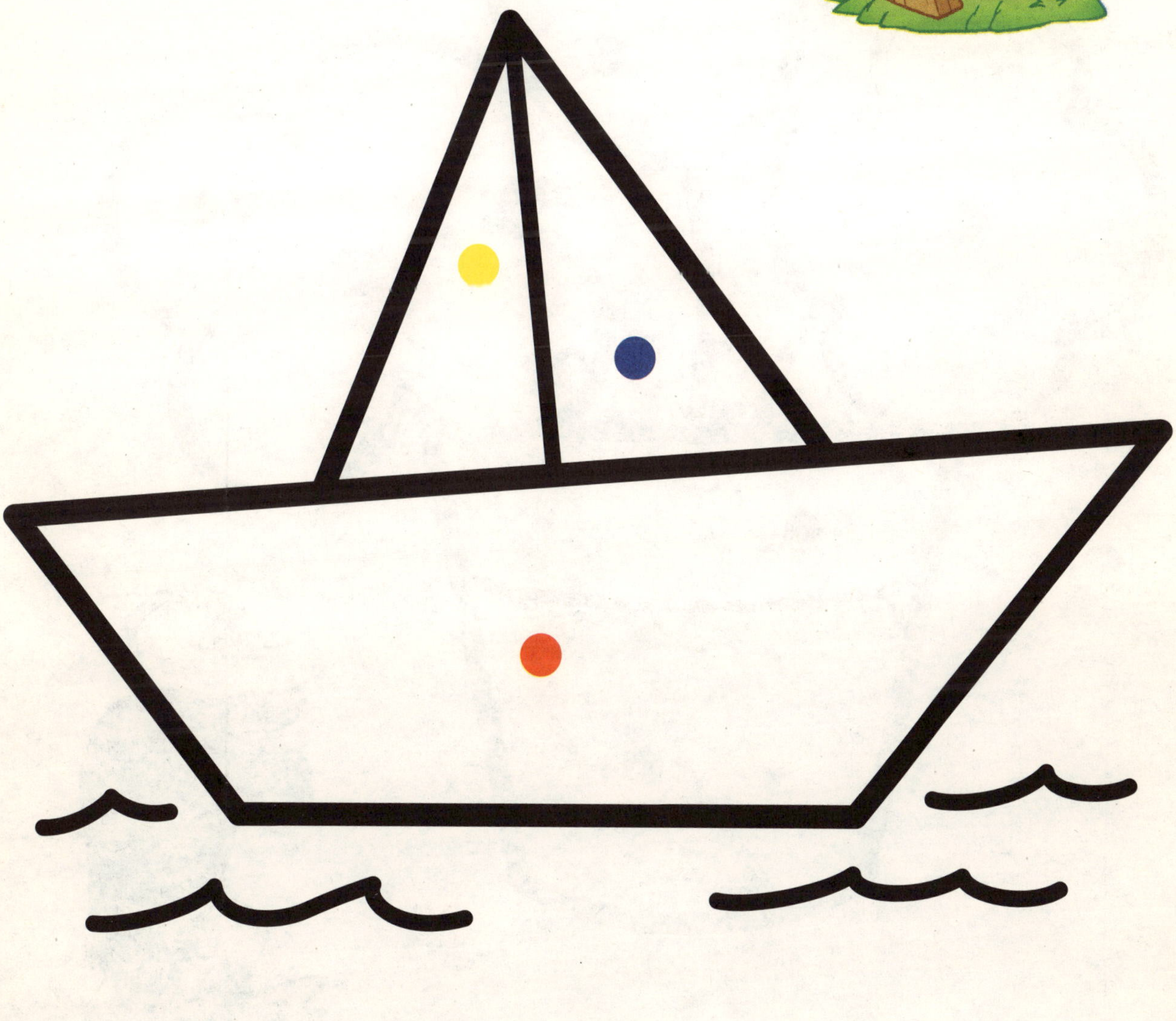

Name: ____________________

Color Cues

Color the drawing with the same colors as the dots.

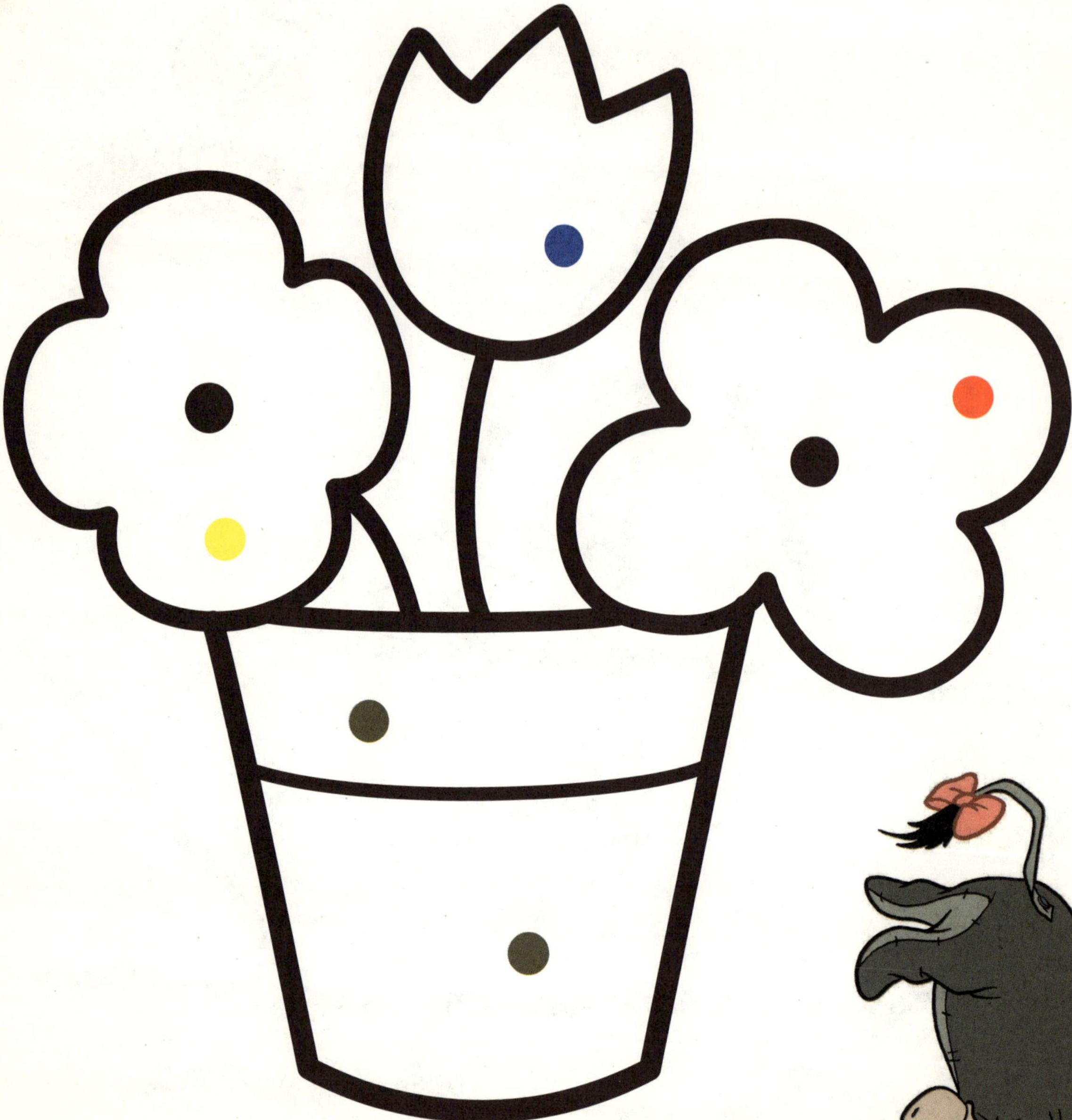

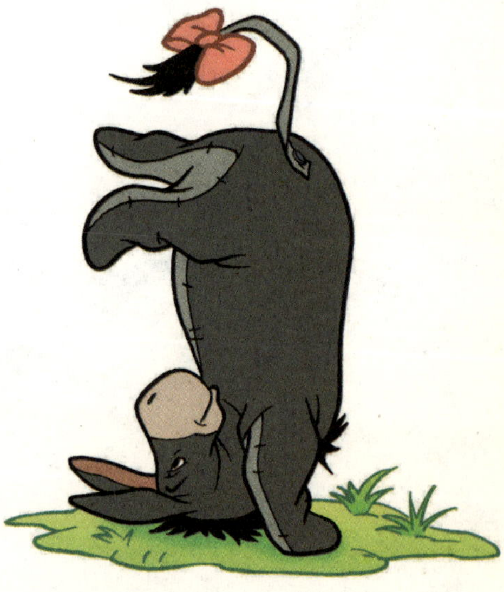

Name: ____________________

Color Cues

Color the drawing with the same colors as the dots.

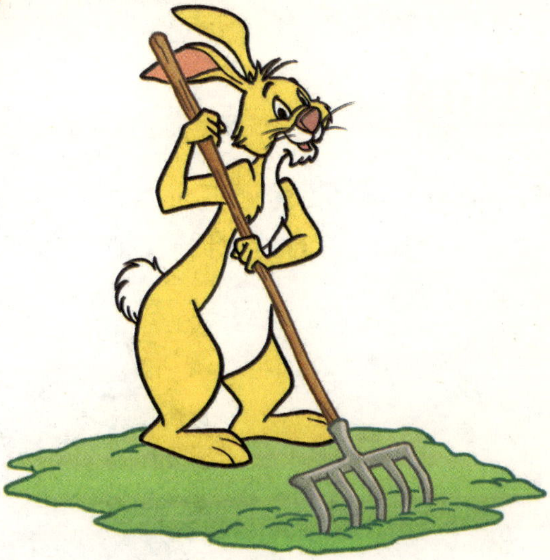

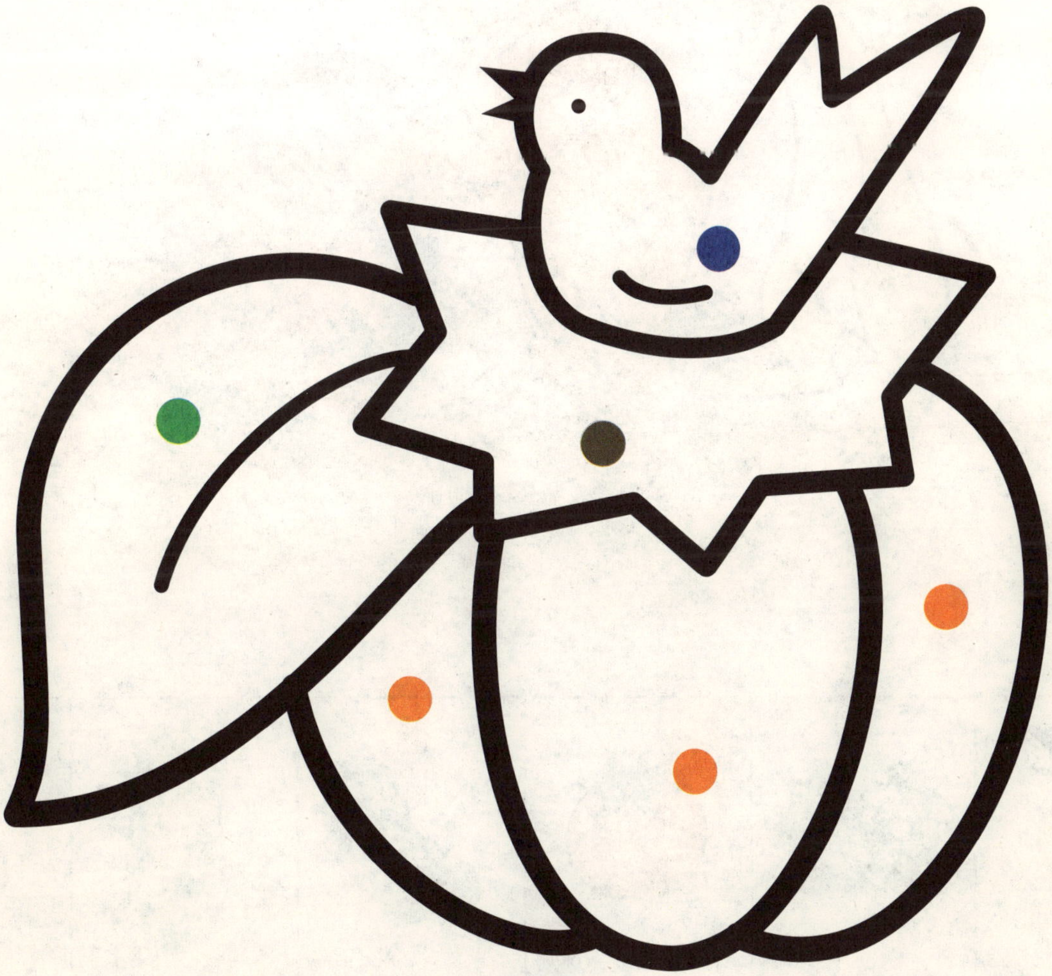

Name: ____________________

Sorting

Help Rabbit and Pooh clean up the vegetables. Have an adult help you cut out each vegetable below and glue it in the correct basket on the next page.

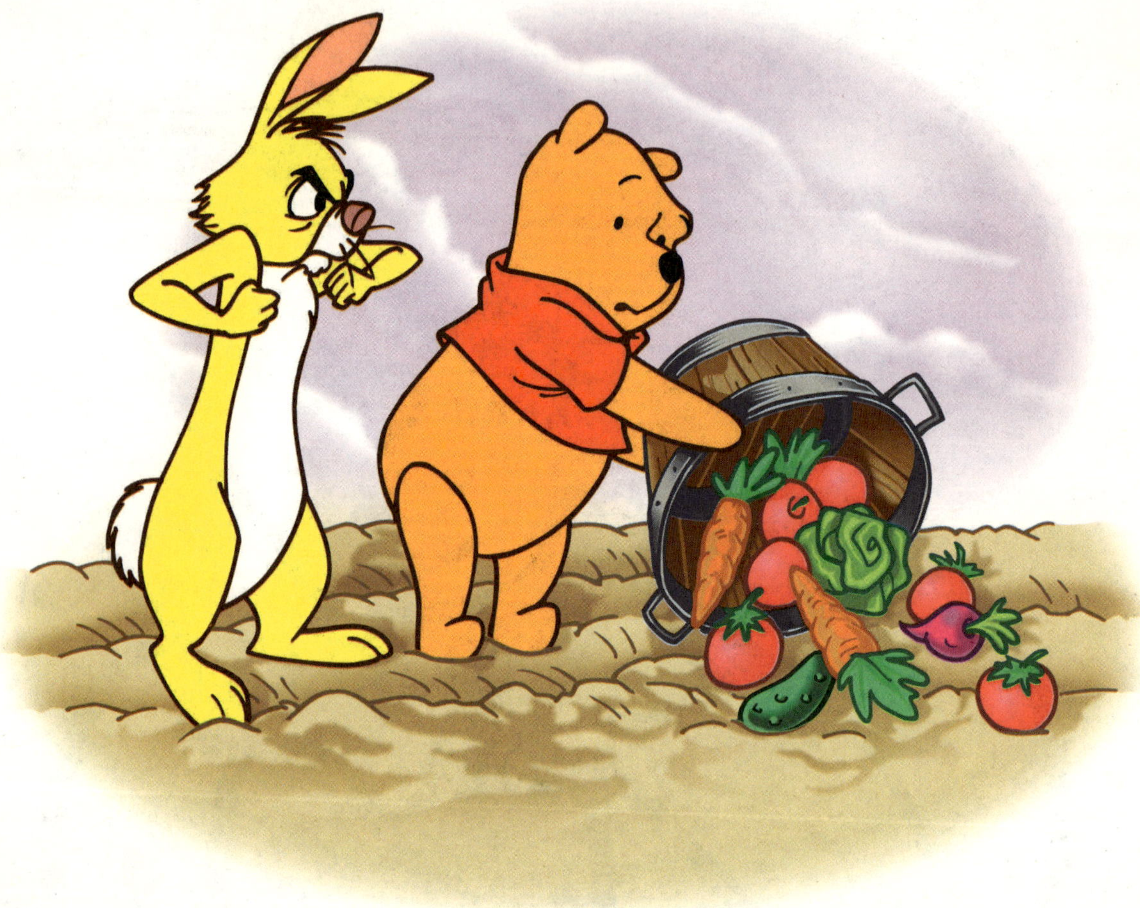

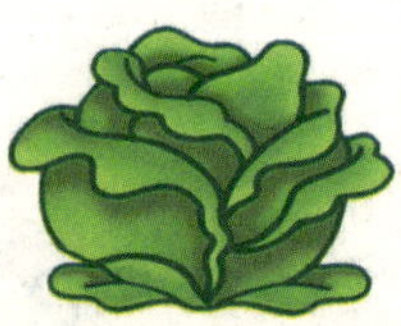

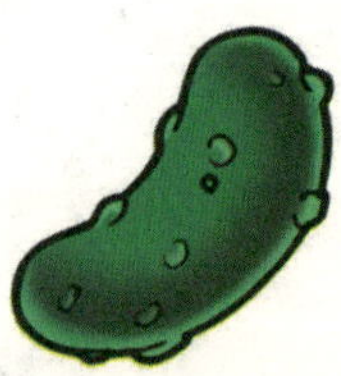

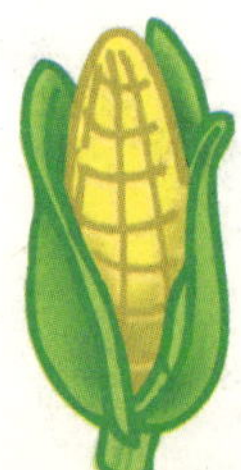

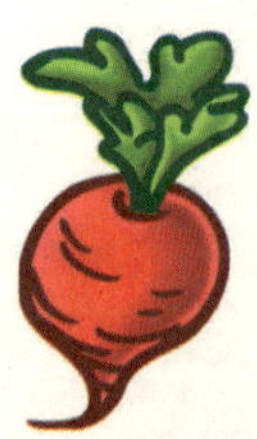

Name: ______________________

Sorting

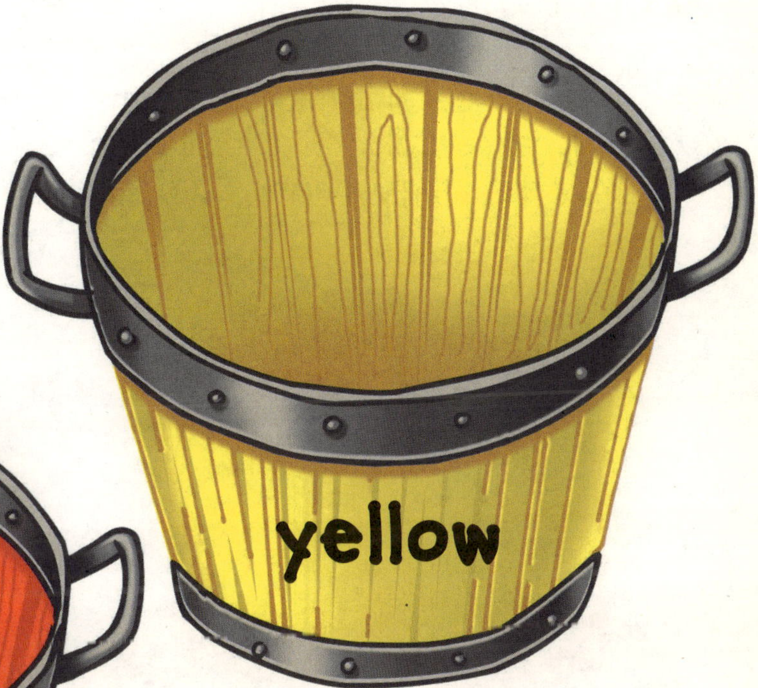

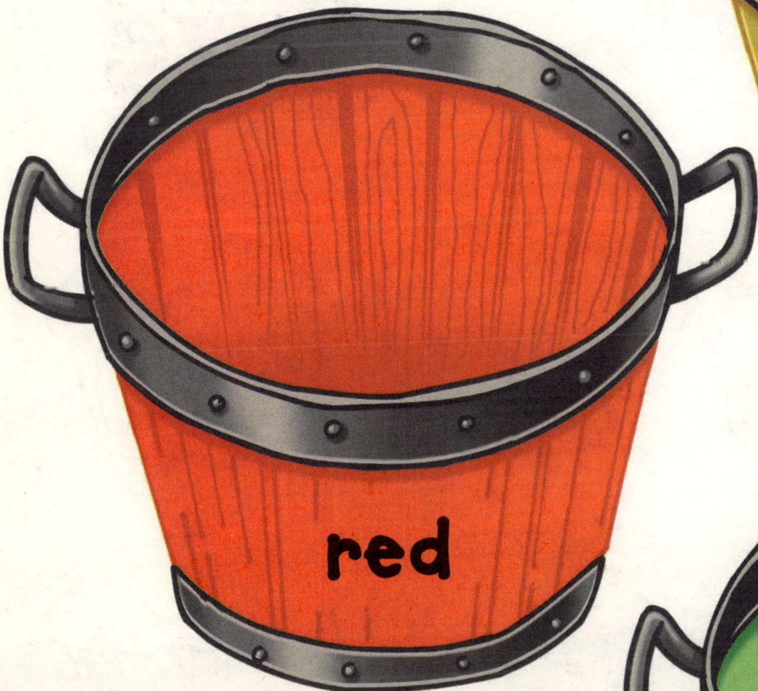

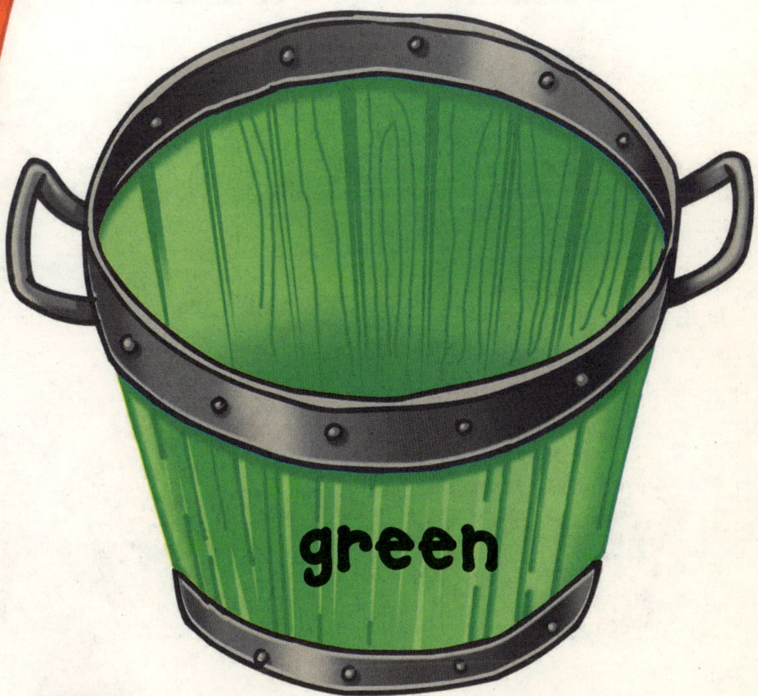

Name: ____________________

Patterns

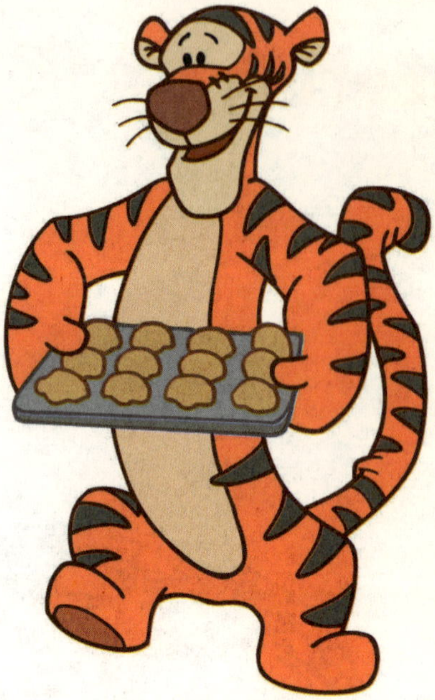

Help Tigger finish decorating the cupcakes. Color the cupcakes to complete the patterns.

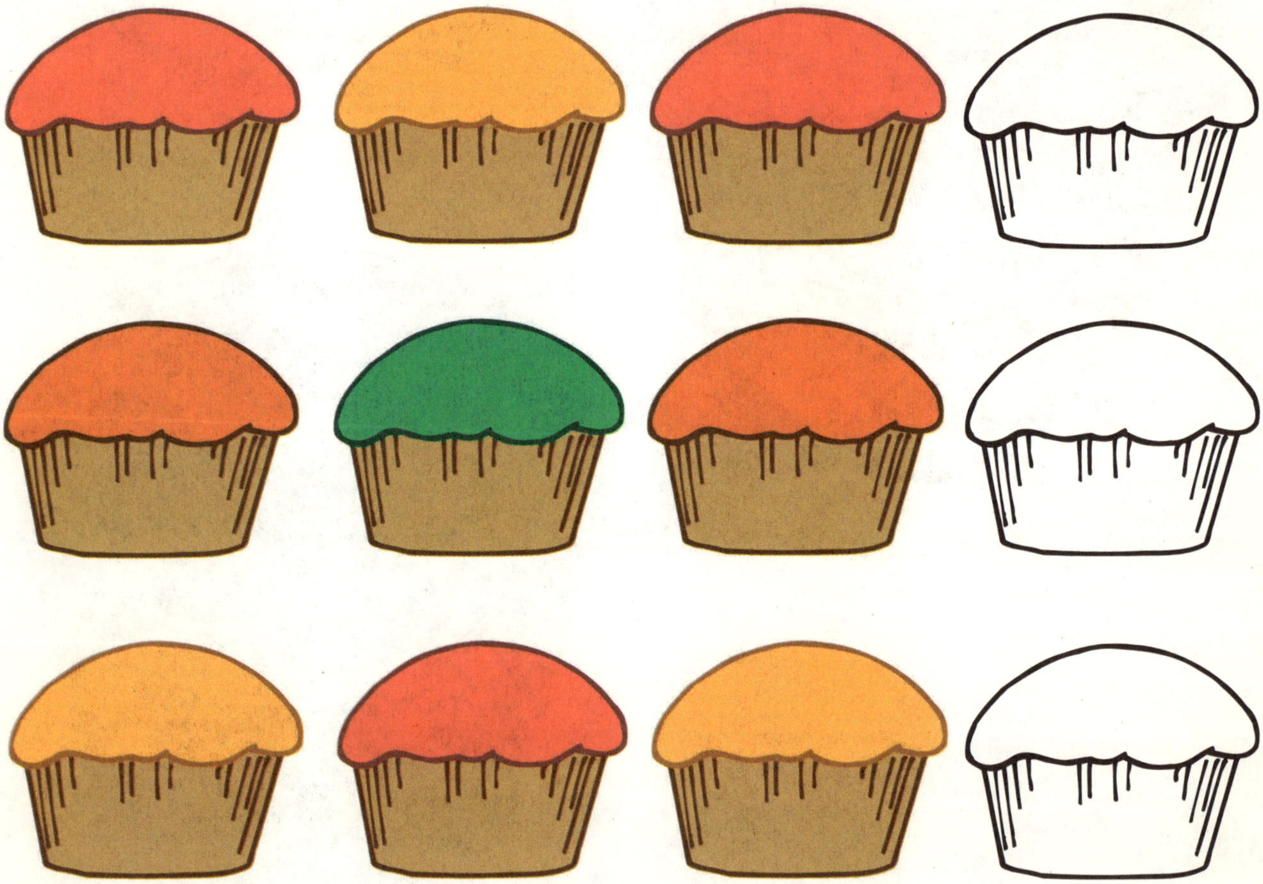

Name: ______________________

Color-by-Number

Help Tigger and Roo color the drawing. Use the colors that match the numbers: 1 (orange); 2 (green); 3 (dark brown).

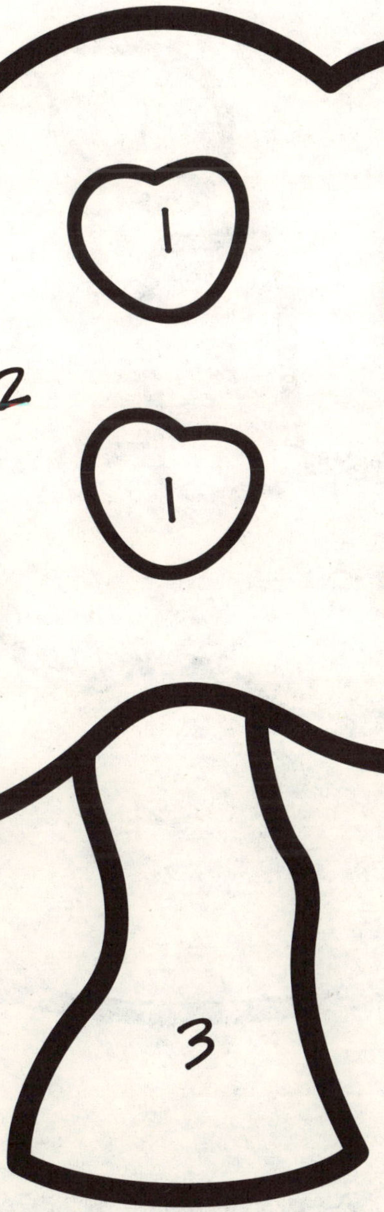

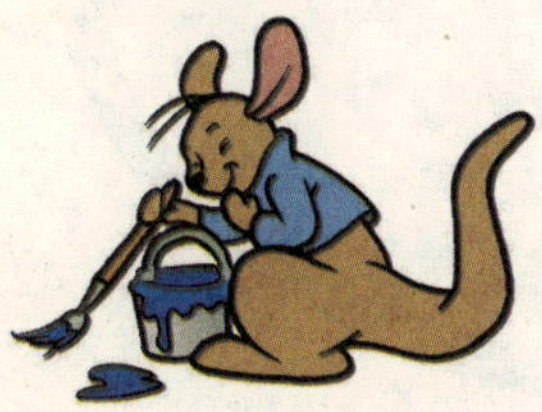

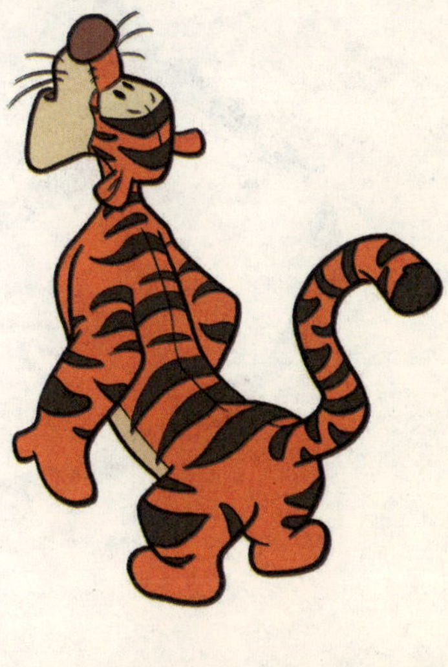

Name: ______________________

Color-by-Number

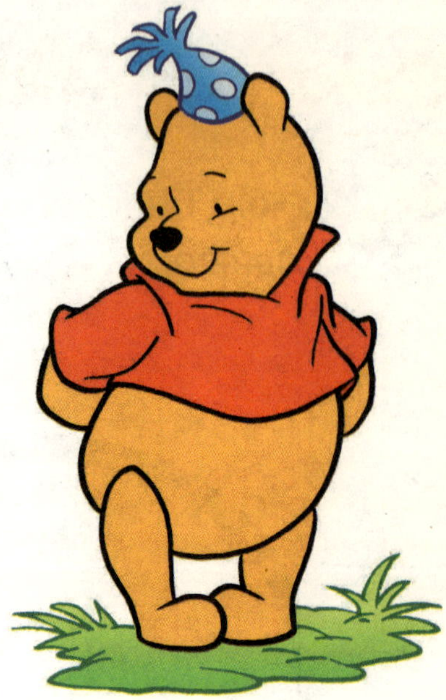

Color Pooh's cake. Use the colors that match the numbers: 1 ●; 2 ●; 3 ●; 4 ●; 5 ●.

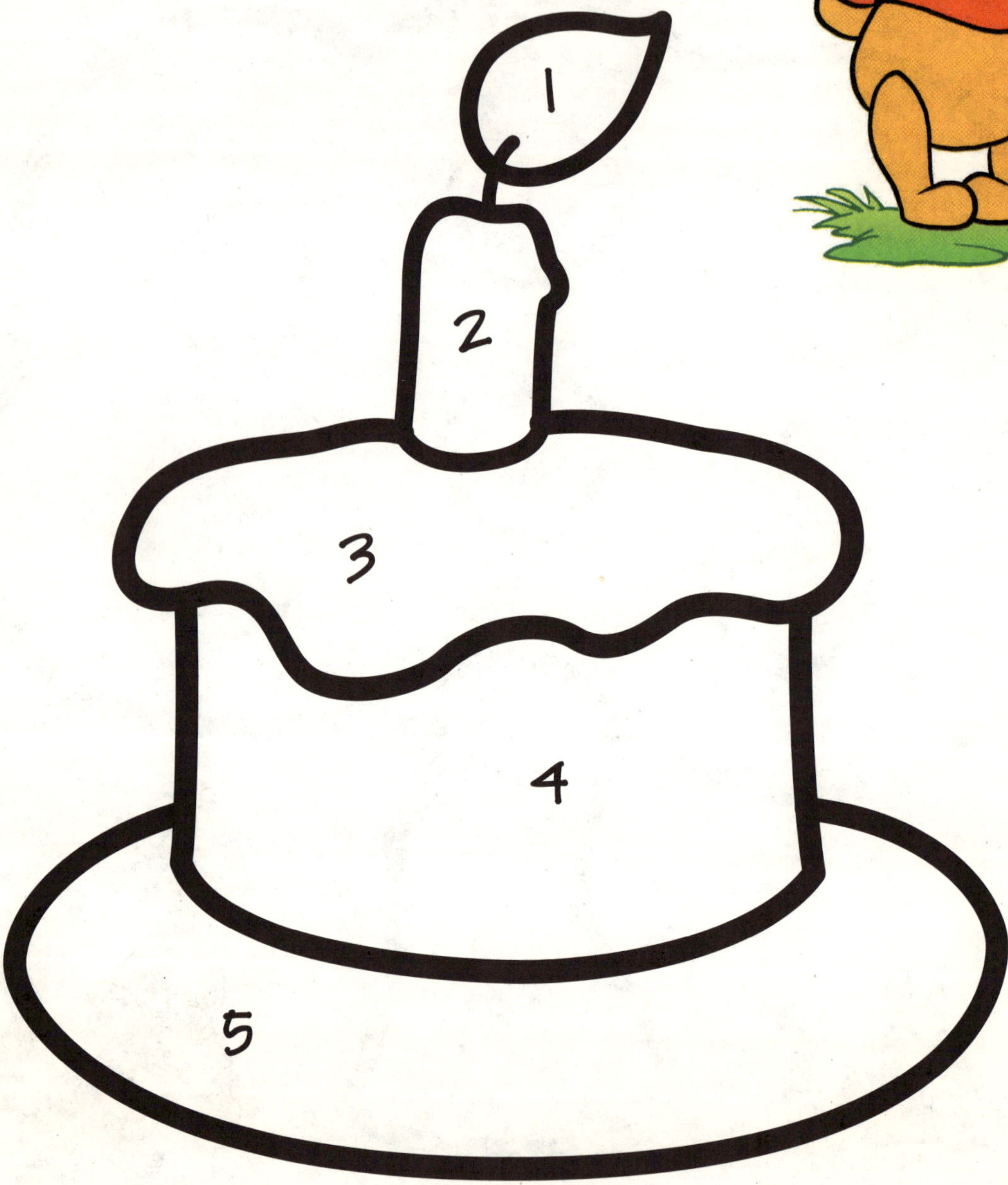

Name: ____________________

Color-by-Number

Color the drawings. Use the colors that match the numbers: 1 ●; 2 ●; 3 ●; 4 ●; 5 ●.

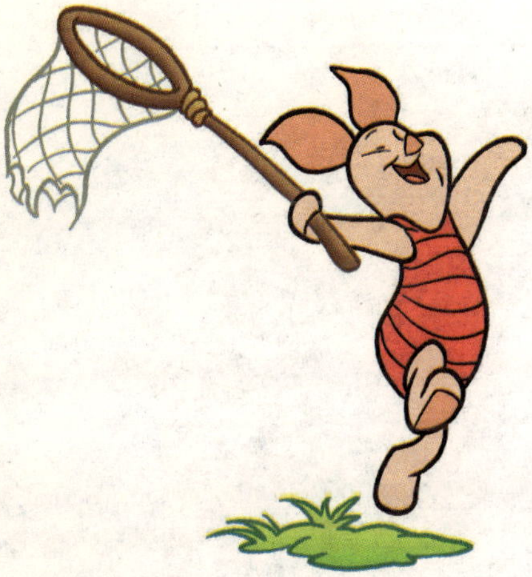

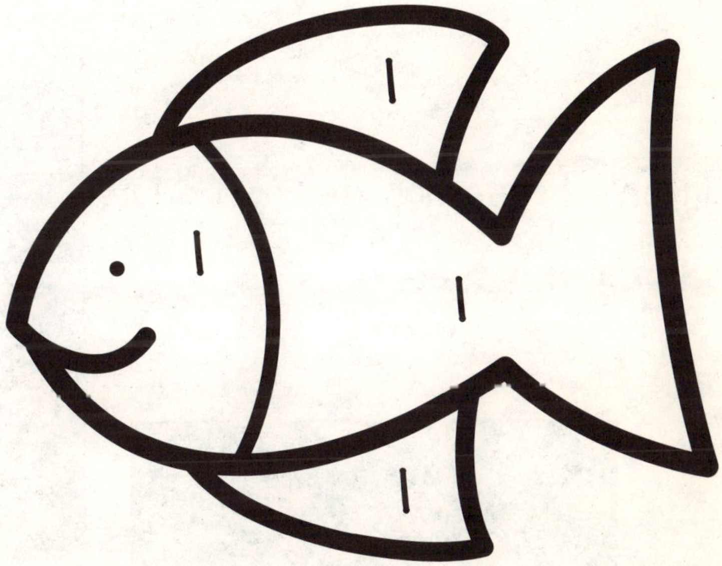

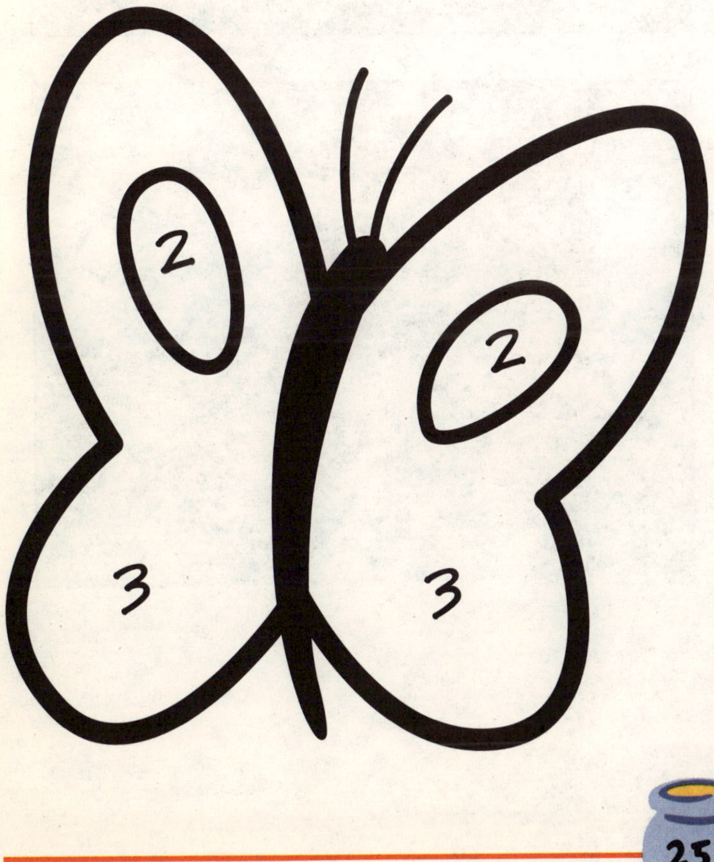

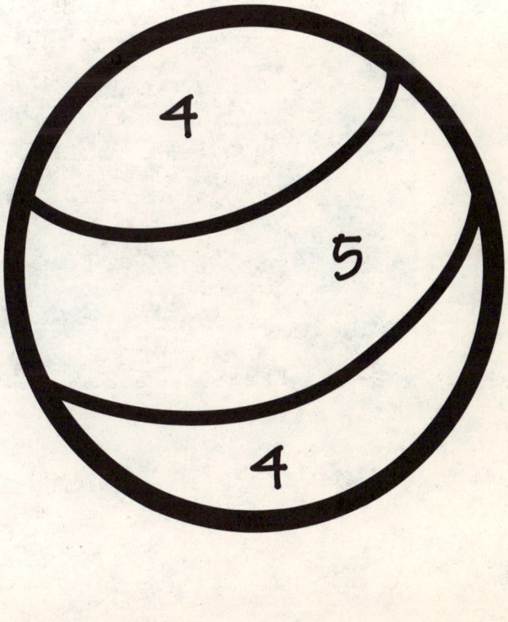

Name: ____________________

Reasoning

Help find Tigger's house. Read the sentences below. Cross out the houses that could not be his. Then, circle Tigger's house.

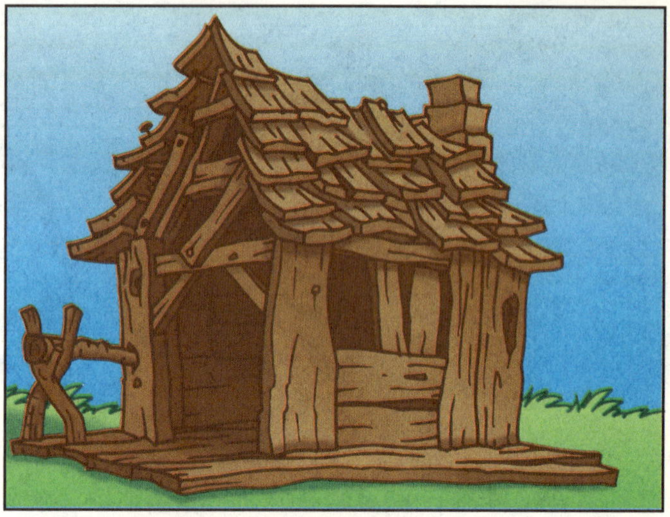

- Tigger's house has a brown door.
- It is up in a tree.

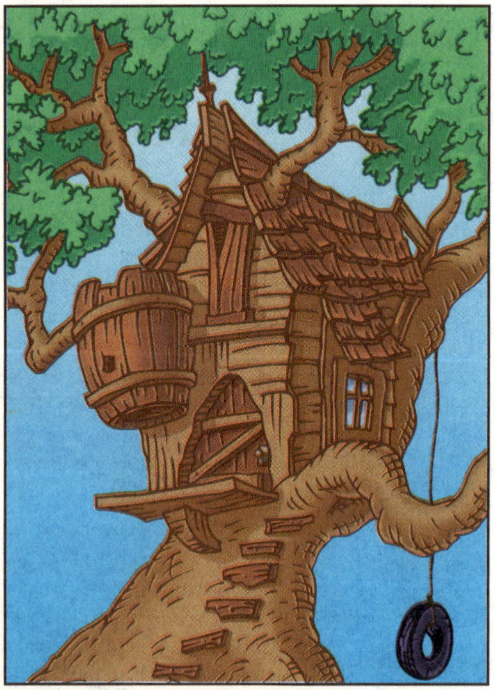

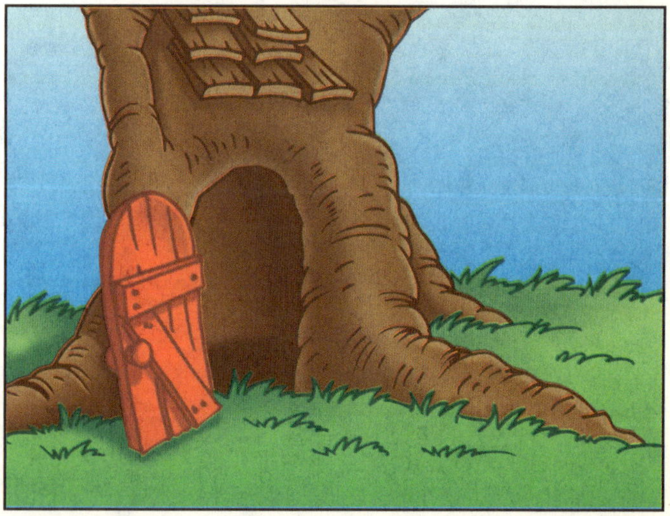

Name: ____________________

Choosing Colors

Color Kanga's table. Use any color you would like.

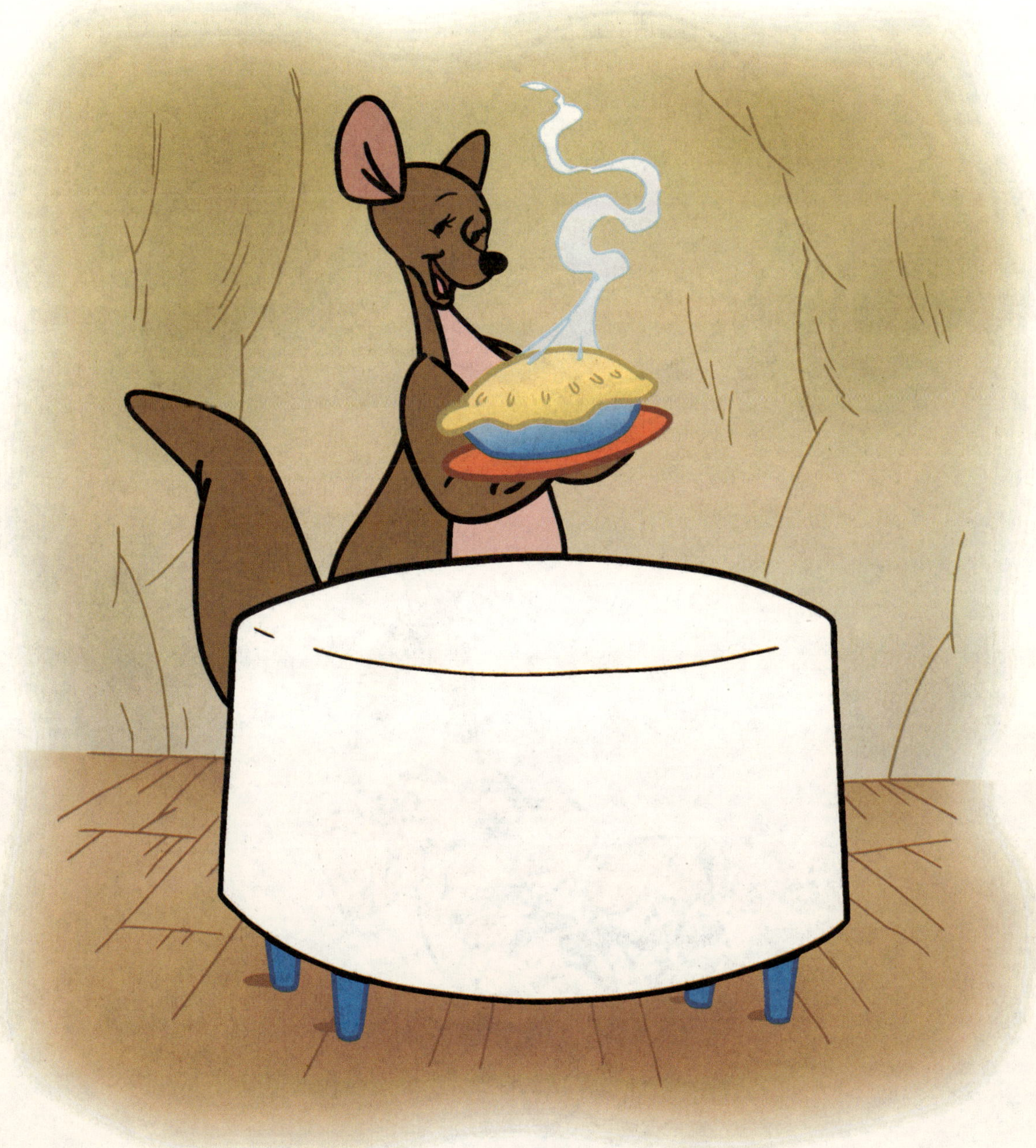

Name: ____________________

Choosing Colors

Color Tigger's balloons. Use any colors you would like.

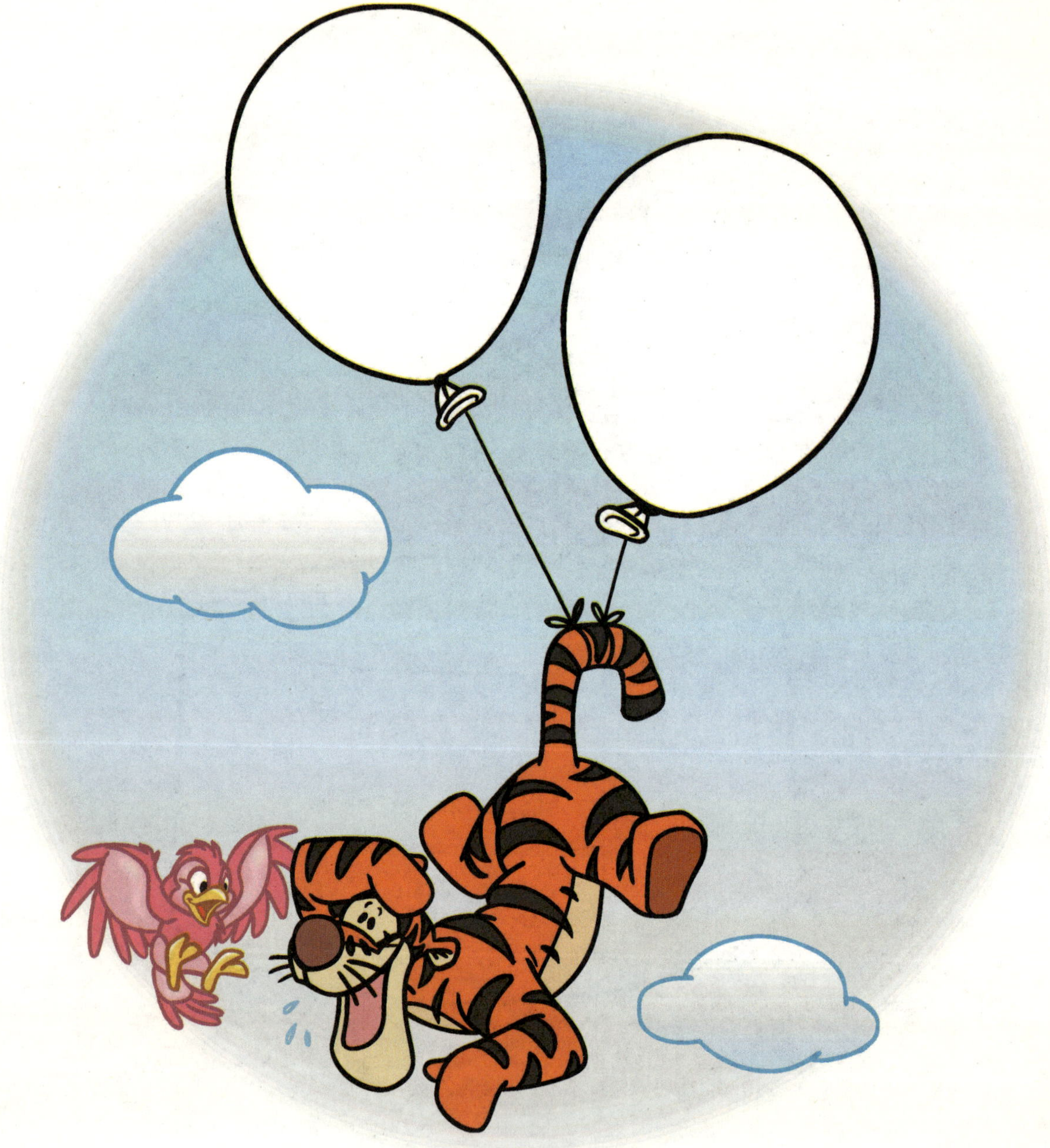

Name: ____________________

Choosing Colors

Color Pooh's blanket and book. Use any colors you would like.

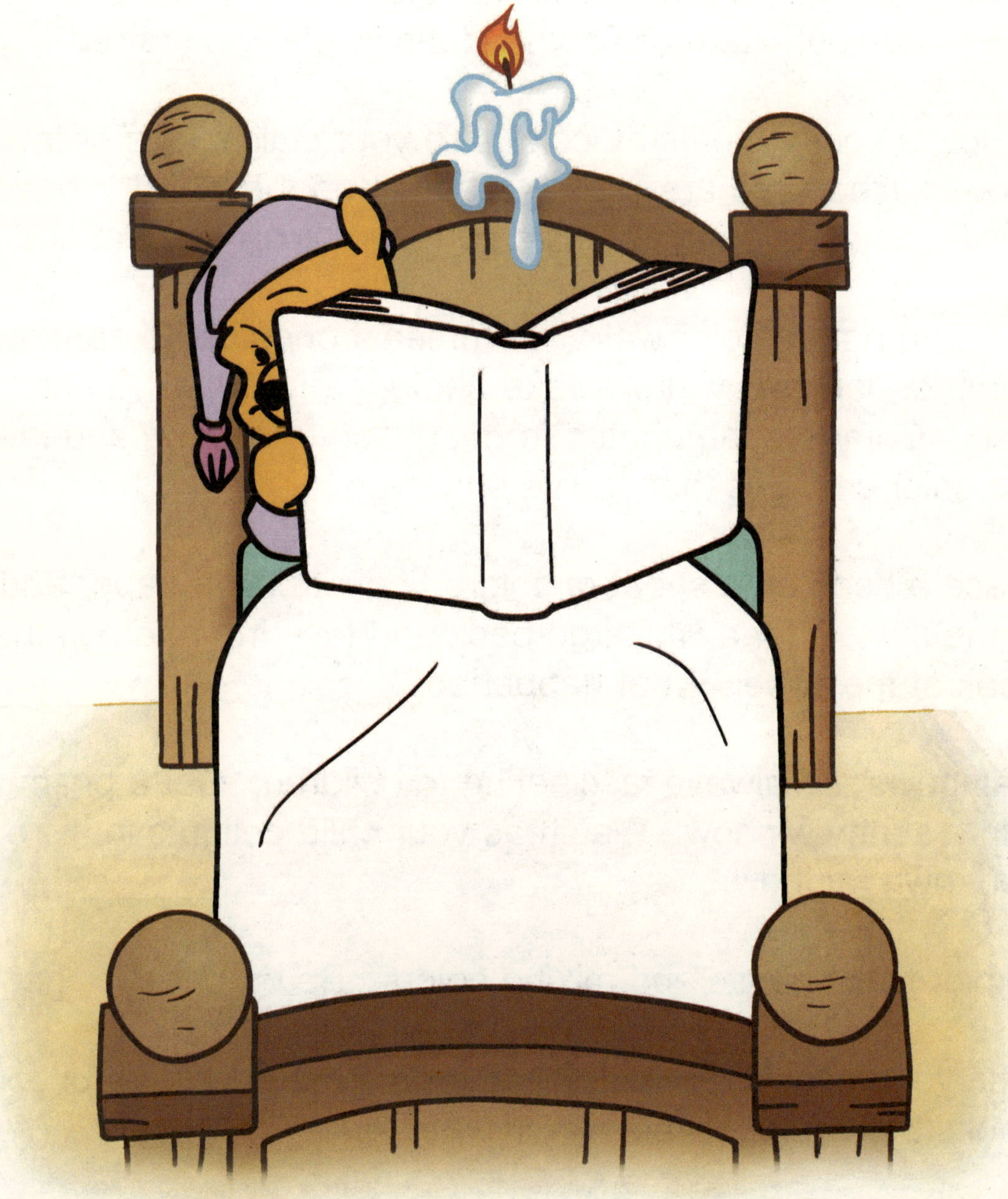

Parent Resource Guide
Helping at Home with Colors

You and your child will have endless hours of fun as you explore the colors all around you! Here are some ideas to get you started:

- Practice sorting items by color, like your child's toys or favorite foods. When you are getting ready to do the laundry, have your child help you sort the clothes by color. What a great skill to learn!

- Go on a nature color walk. Go at least once in the spring, summer, fall, and winter. Take photographs while you are on your walk. Compare the photos from season to season and talk about the color changes that you've observed.

- Place a book on a sheet of colored construction paper and put it in the sun. After a while, go back and take the book off the paper. Look at the paper. What happened?

- Rainbows are always fascinating to children. Get a prism and put it in a sunny window. Watch as your child delights in his or her very own rainbow.

- Explore with paints and mixing colors. Be prepared to get messy!

- Visit your local library and discover the many wonderful color books that are available.

Page 11.

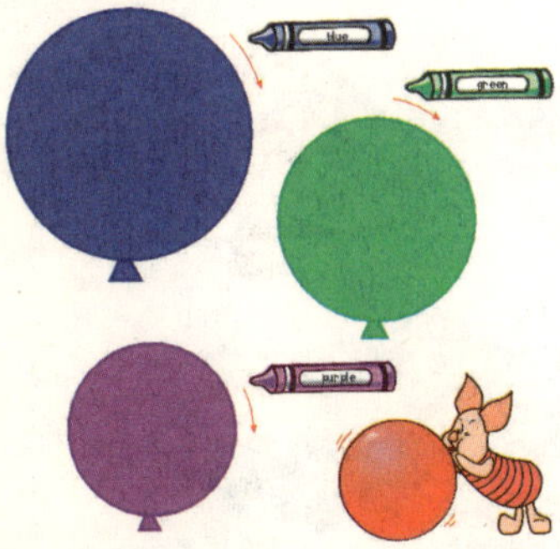

Page 14.

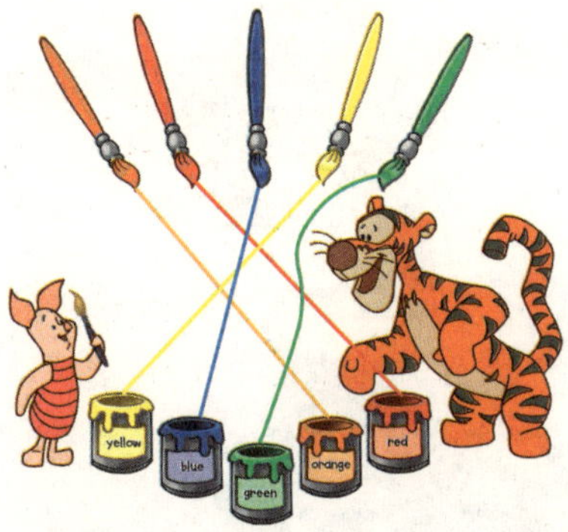

Page 15.

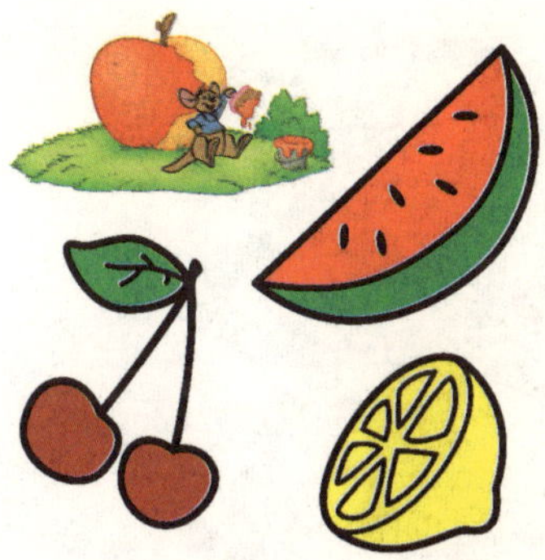

Page 16.

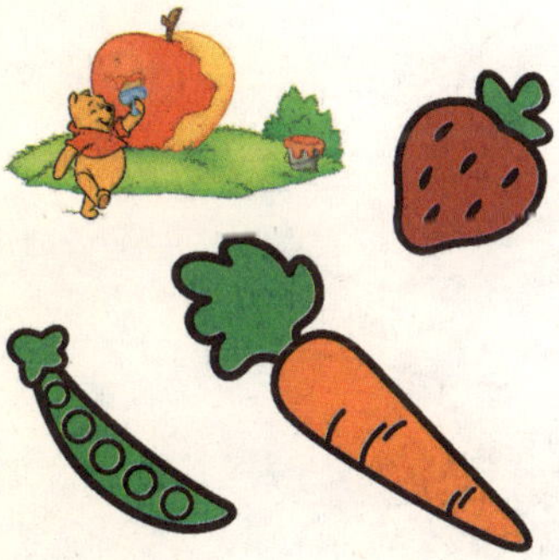

Page 17.

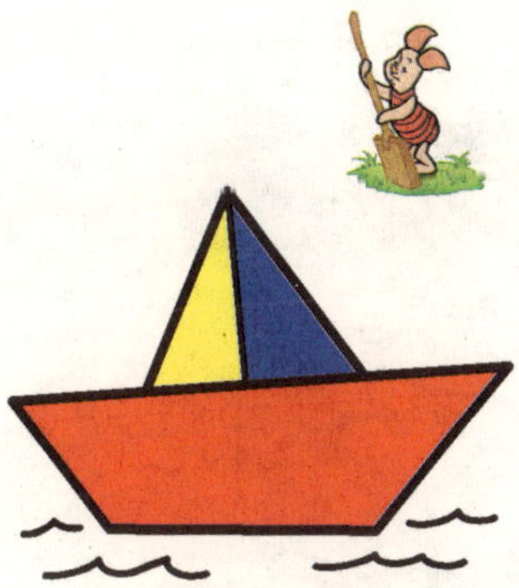

Page 18.

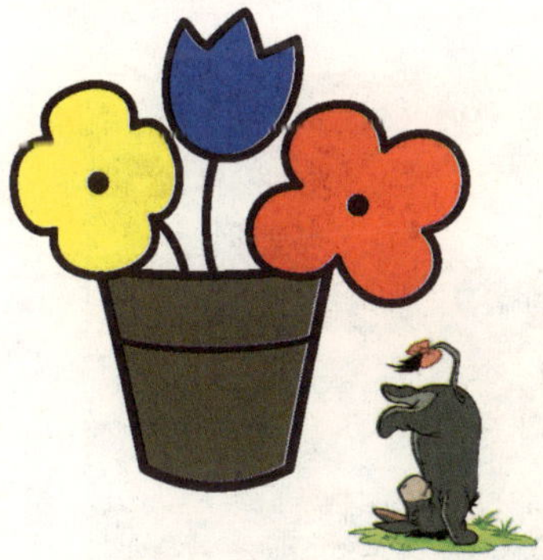

Page 19.

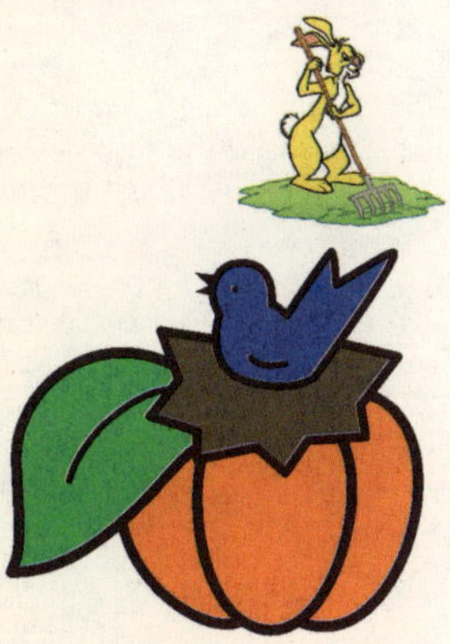

Page 21.

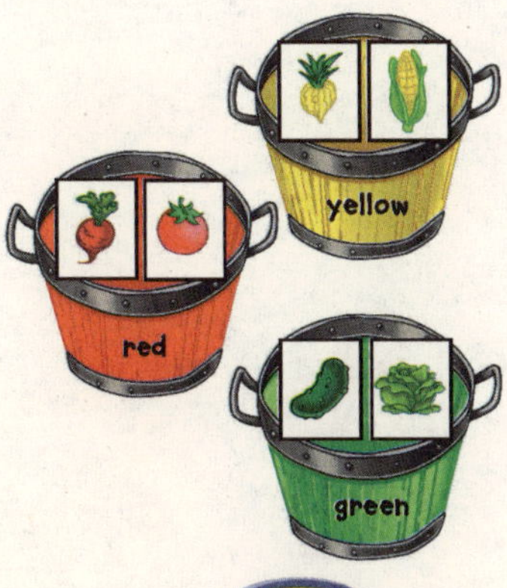

Page 22.

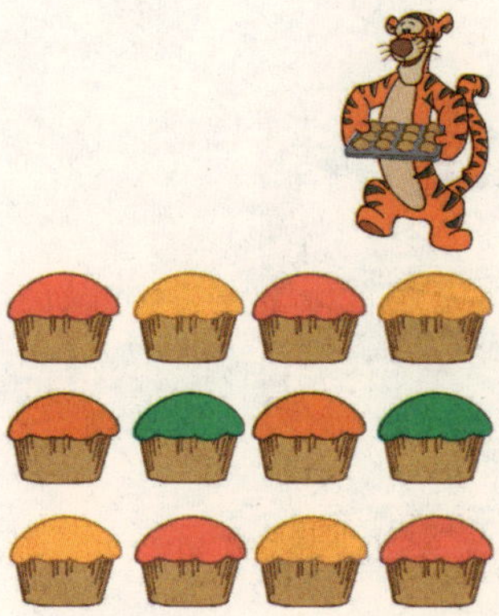

Page 23.

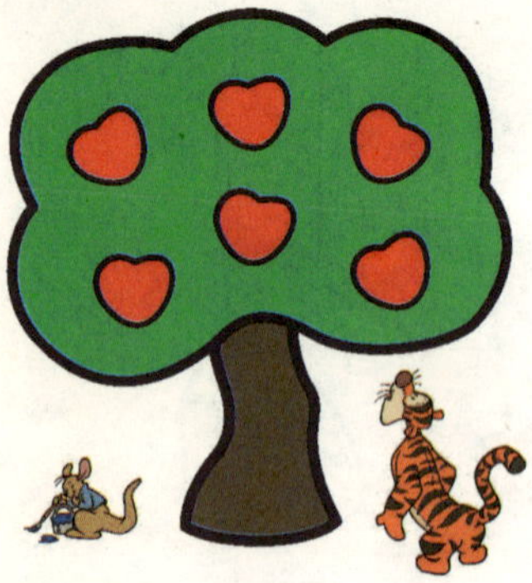

Page 24.

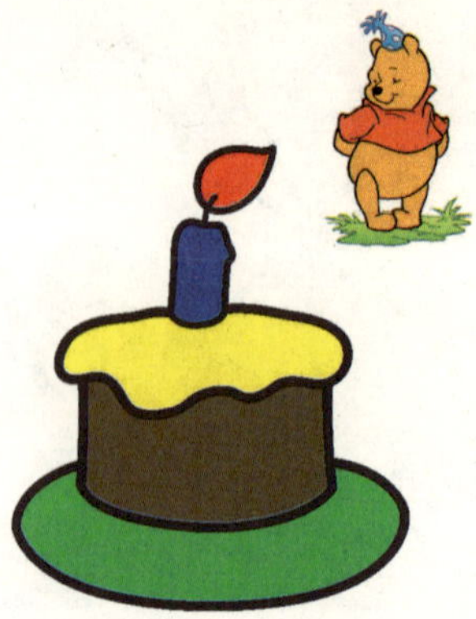

Page 25.

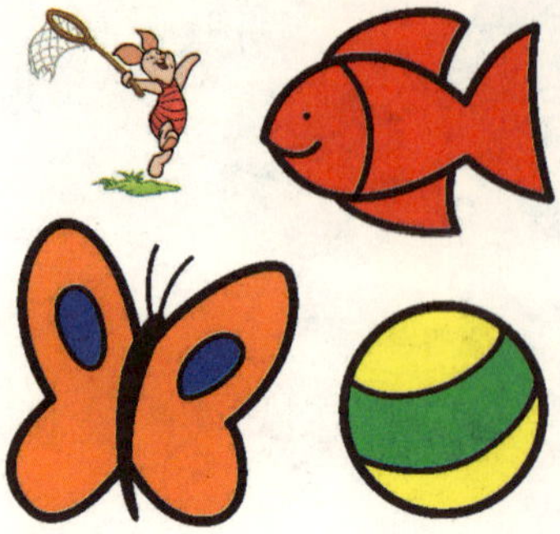

Page 26.

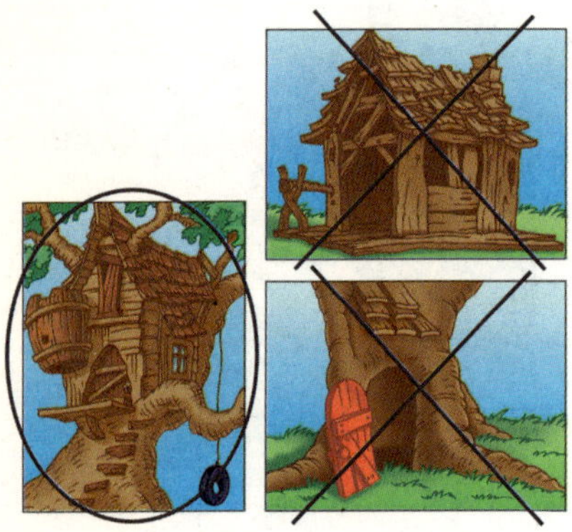

Page 27.

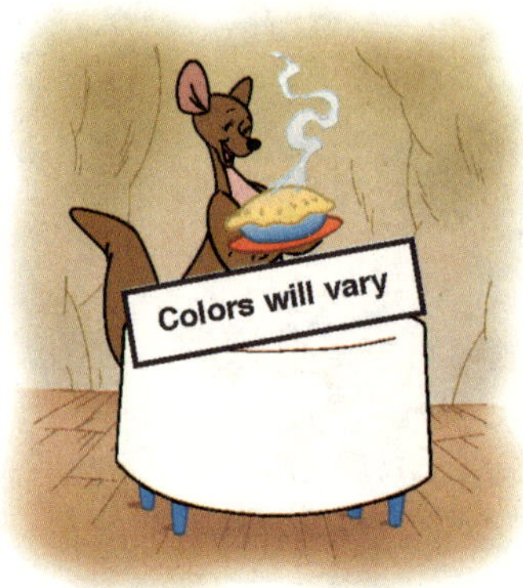

Page 28.

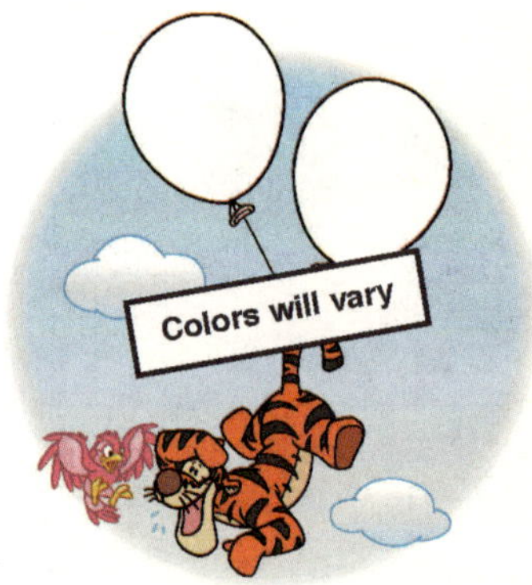

Page 29.

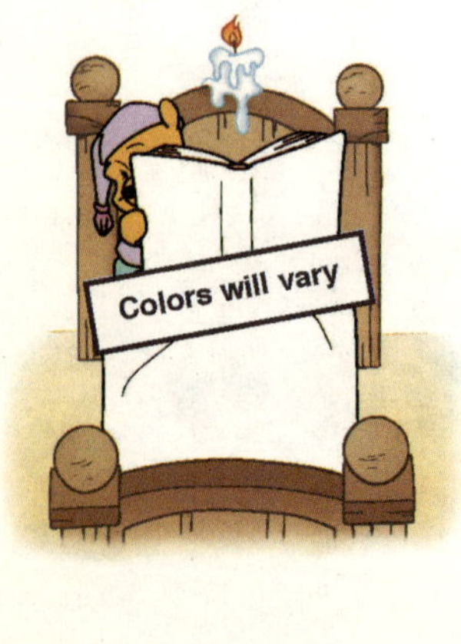